EAI/Springer Innovations in Communication and Computing

Series Editor

Imrich Chlamtac, European Alliance for Innovation, Ghent, Belgium

The impact of information technologies is creating a new world yet not fully understood. The extent and speed of economic, life style and social changes already perceived in everyday life is hard to estimate without understanding the technological driving forces behind it. This series presents contributed volumes featuring the latest research and development in the various information engineering technologies that play a key role in this process. The range of topics, focusing primarily on communications and computing engineering include, but are not limited to, wireless networks; mobile communication; design and learning; gaming; interaction; e-health and pervasive healthcare; energy management; smart grids; internet of things; cognitive radio networks; computation; cloud computing; ubiquitous connectivity, and in mode general smart living, smart cities, Internet of Things and more. The series publishes a combination of expanded papers selected from hosted and sponsored European Alliance for Innovation (EAI) conferences that present cutting edge, global research as well as provide new perspectives on traditional related engineering fields. This content, complemented with open calls for contribution of book titles and individual chapters, together maintain Springer's and EAI's high standards of academic excellence. The audience for the books consists of researchers, industry professionals, advanced level students as well as practitioners in related fields of activity include information and communication specialists, security experts, economists, urban planners, doctors, and in general representatives in all those walks of life affected ad contributing to the information revolution.

Indexing: This series is indexed in Scopus, Ei Compendex, and zbMATH.

About EAI - EAI is a grassroots member organization initiated through cooperation between businesses, public, private and government organizations to address the global challenges of Europe's future competitiveness and link the European Research community with its counterparts around the globe. EAI reaches out to hundreds of thousands of individual subscribers on all continents and collaborates with an institutional member base including Fortune 500 companies, government organizations, and educational institutions, provide a free research and innovation platform. Through its open free membership model EAI promotes a new research and innovation culture based on collaboration, connectivity and recognition of excellence by community.

Francisco Torres-Guerrero • Leticia Neira-Tovar •
Jorge Bacca-Acosta
Editors

2nd EAI International Conference on Smart Technology

Springer

EAI
RESEARCH MEETS INNOVATION

Editors
Francisco Torres-Guerrero
Universidad Autónoma de Nuevo León
Ciudad, Mexico

Leticia Neira-Tovar
Universidad Autónoma de Nuevo León
San Nicolás de los Garza, Mexico

Jorge Bacca-Acosta
Fundación Universitaria Konrad Lorenz
Bogotá, Colombia

ISSN 2522-8595 ISSN 2522-8609 (electronic)
EAI/Springer Innovations in Communication and Computing
ISBN 978-3-031-07669-5 ISBN 978-3-031-07670-1 (eBook)
https://doi.org/10.1007/978-3-031-07670-1

This Springer imprint is published by the registered company Springer Nature Switzerland AG
The registered company address is: Gewerbestrasse 11, 6330 Cham, Switzerland

Preface

We are delighted to introduce the proceedings of the second edition of the 2021 European Alliance for Innovation (EAI) International Conference EAI MTYMEX 2021 – 2nd EAI International Conference on Smart Technology. This conference has brought together researchers, developers, and practitioners around the world who are leveraging and developing smart grid technology for a smarter and more resilient grid.

The EAI MTYMEX is an international scientific conference to demonstrate the potential of new applications for the Internet in the future (American and European universities will be participating). The conference brought together the academic, research, and private sectors working on applications and smart devices for e-health and smart cities.

The technical program consisted of 12 full papers. The conference tracks were: Track 1 – Usage of Smart Technology in Healthcare; Track 2 – Usage of Smart Technology for Smart Society; Track 3 – Usage of Smart Technology – FI-Artificial Intelligence; and Track 4 – Usage of Smart Technology in Information Security. Apart from the high-quality technical paper presentations, the technical program also featured three keynote speeches. The first keynote speaker was Dr. Nasser Mohamed, Director of Technological Innovation at the UANL and Liaison of the UANL for Industry 4.0 affairs. The title of his speech was "Digitization and Quality: An Effective Pairing for Smart Industry." The second keynote speaker was Dr. José Abreu with the speech "Social Responsibility of the Pharmaceutical Industry and Its Stakeholders in Times of Coronavirus." The third keynote speaker was Paulino Calderon with the speech "Practical IoT Hacking" – co-funding Websec México.

Coordination with the steering chair Imrich Chlamtac was essential for the success of the conference. We sincerely appreciate their constant support and guidance. It was also a great pleasure to work with such an excellent organizing committee team for their hard work in organizing and supporting the conference, in particular the Technical Program Committee, led by Victor Leyva. We are also grateful to Conference Manager Aleksandra Sledziejowska for her support and all the authors who submitted their papers to the EAI MTYMEX 2021 conference.

We strongly believe that EAI MTYMEX 2021 conference provides a good forum for all researcher, developers, and practitioners to discuss all science and technology aspects that are relevant to smart cities. We also expect that the future conference will be as successful and stimulating, as indicated by the contributions presented in this volume.

Ciudad, Mexico
San Nicolás de los Garza, Mexico
Bogotá, Colombia

Francisco Torres-Guerero
Leticia Neira-Tovar
Jorge Bacca-Acosta

Conference Organization

Steering Committee
Imrich Chlamtac
Bruno Kessler Professor, University of Trento, Italy

Organizing Committee
General Chair
Francisco Torres Guerrero
The Autonomous University of Nuevo León, Mexico

Leticia Neira
The Autonomous University of Nuevo León, Mexico

TPC Chair and Co-Chair
Victor Leyva
The Autonomous University of Nuevo León, Mexico

Sponsorship and Exhibit Chair
Nasser Noriega
The Autonomous University of Nuevo León, Mexico

Local Chair
Fernando Banda
The Autonomous University of Nuevo León, Mexico

Workshops Chair
Salvador Barrera
Universidad Regiomontana, Mexico

Publicity & Social Media Chair
Oscar Rangel
The Autonomous University of Nuevo León, Mexico

Publications Chair
Romeo Sanchez
The Autonomous University of Nuevo León, Mexico

Web Chair
Sergio Ordoñez
The Autonomous University of Nuevo León, Mexico

Posters and PhD Track Chair
Carlos Chavez
The Autonomous University of Nuevo León, Mexico

Panels Chair
Alfonso Lopez
The Autonomous University of Nuevo León, Mexico

Demos Chair
Jose Luis Abreau
SPENTA University, Mexico

Tutorials Chairs
Mayra Flores
The Autonomous University of Nuevo León, Mexico

Reception Chair
Arnulfo Cubero
The Autonomous University of Nuevo León, Mexico

Contents

A Location-Based Mobile Advertising System for Small-to-Medium Businesses

Ahmed Abdelmoamen Ahmed and Anitha Palusa

Abstract Business owners need more affordable venues to proclaim their services and products. Mobile technologies offer a convenient path for implementing smart location-based advertising (LBA) solutions using intelligent handheld devices. This chapter presents a case study for a mobile-based LBA system that has components running on smartphones as a mobile app and cloud servers as a web application. Google Maps is used in the mobile app as an underlying service for enhancing the experience of mobile users in using the system. Mobile users are able to utilize the mobile app to travel to and from their destinations meanwhile seeing and engaging relevant advertisements, including new job openings located in their local neighborhoods and discounts on their favorite meals. This chapter proposes a novel sensing approach for representing the switching of sensing contexts in the proposed LBA system. The main goal of multi-modal sensing is to decrease the energy consumed from the mobile app by identifying the changes in the current sensing mode automatically, thus avoiding needless sensing overheads. We organized the proposed LBA system into these two components: mobile user and business owner sides. First, the business owners are allowed to initiate new advertisements by entering simple metadata about their advertisements on the business side. Second, mobile users, on the user side, can search the Google Map to see attractive advertisements while doing their daily activities such as driving, bicycling, jogging, walking, or relaxing in their homes. We carried out a set of experimental evaluations to show the performance and scalability of the proposed LBA approach.

Keywords LBA · Multi-modal sensing · Mobile app · Energy-efficient

A. A. Ahmed (✉) · A. Palusa
Computer Science Department, Prairie View A&M University, Prairie View, TX, USA
e-mail: amahmed@pvamu.edu

F. Torres-Guerrero et al. (eds.), *2nd EAI International Conference on Smart Technology*, EAI/Springer Innovations in Communication and Computing,
https://doi.org/10.1007/978-3-031-07670-1_1

1 Introduction

Mobile and intelligent technologies have created an opportunity for small businesses to advertise their products affordably using LBA technologies.Using LBA, the owners of small businesses can customize their ads to potential customers in real time according to their contemporary geolocation. This has taken away the constraints between customers and businesses when they are in close proximity to the target business locations [14].

In this chapter, we present a prototype implementation of a mobile-based LBA solution that various types of businesses can use to promote their services and products in an affordable way. The developed system is organized into components running on mobile devices and a cloud server as a web-based application. We developed a mobile app on the user side to increase the LBA system's user experience. We developed a web application on the business side, allowing business owners to create and manage various types of advertisements via an interactive GUI.

The power efficiency of smartphones becomes a crucial factor in developing mobile apps. In this chapter, we found an opportunity for conserving the energy of the developed mobile app by optimizing the sensing process of mobile devices by modeling its evolving sensing context. We call this approach *context-aware multi-modal sensing* [1]. Utilizing multi-modal sensing, we could represent the evolving sensing needs of mobile apps that can change their behavior based on the changes in context that a mobile device is in real time.

We conducted experimental evaluations on the scalability and power demand when using our LBA solution. The experimental results have demonstrated the energy efficiency and scalability of the proposed multi-modal system.

There are many interesting projects—in academia (e.g., [17, 38]) and industry (e.g., [21, 37])—that involve LBA applications and services in different domains such as monitoring traffic data [20, 37], rescue management [15], preventive health [34, 39], games and entertainment [35], crowdsensing [19], and mobile-based advertising [4, 33].

The proposed approach is more related to research focused on providing support for mobile-based LBA frameworks. The existing work has taken different ways to support LBA applications and services, which focus on the programmability [10] and participatory crowdsensing [22].

Compared to the existing work, many existing platforms for LBA systems are either expensive products or periodically paid services that small businesses cannot afford. Moreover, the existing solutions are designed to study the attitudes of mobile users [14] or analyze the existing business models [18]. Furthermore, they cannot support the simultaneous execution of many LBA services on a single platform, eliminating the opportunities to optimize the overall power consumption by sharing similar sensing requirements among these LBA services [1].

2 Design

As shown in Fig. 1, the proposed LBA architecture is organized into components running on cloud servers (i.e., application and database) and mobile devices such as smartphones and tablets. We used the REST service to coordinate the communication between the two sides.

The proposed multi-modal approach—for modeling the sensing context switching—is implemented at the user side as a mobile app. The multi-modal sensing process could be represented by using a finite-state machine (FSM), which we model as follows:

$$\langle M, \Sigma, \delta, M_0 \rangle, \tag{1}$$

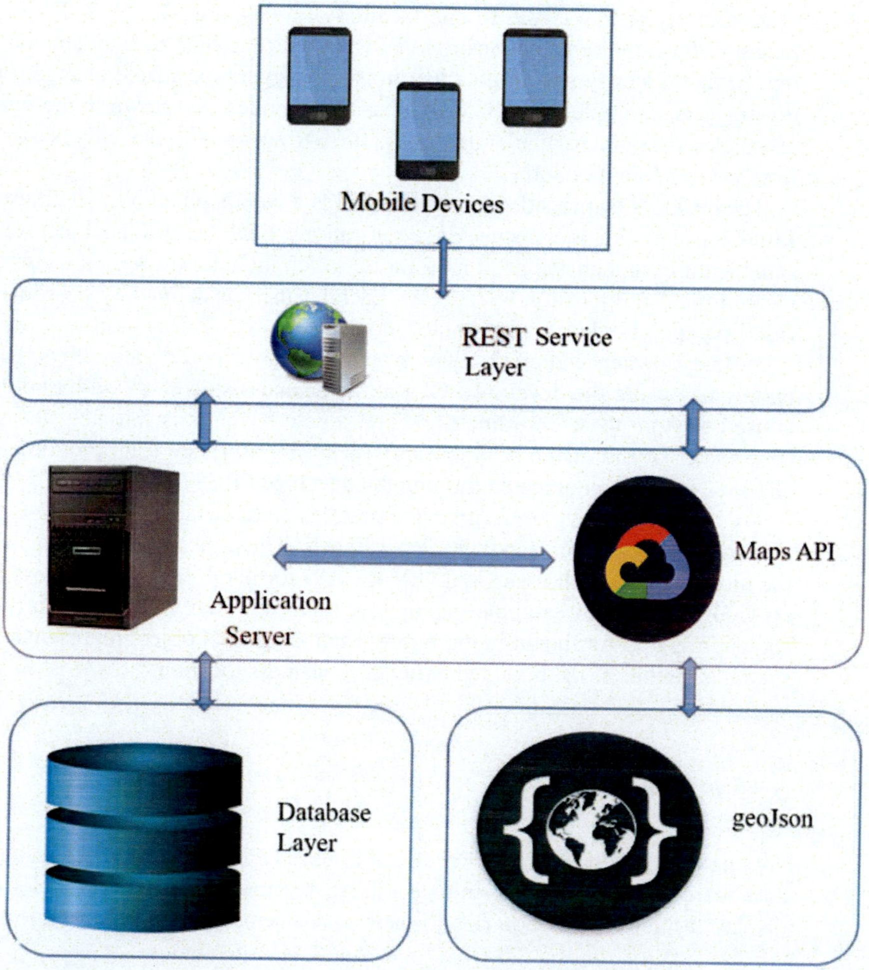

Fig. 1 System architecture

Fig. 2 Multi-modal sensing

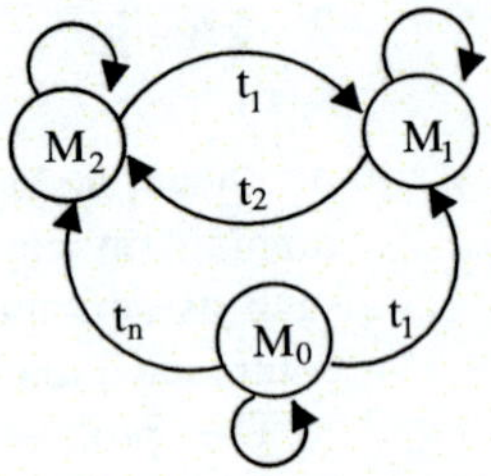

where M is a non-empty finite set of sensing modes, Σ is a non-empty finite set of the input sensing data, and $M_0 \in M$ is the initial sensing state.

Every mode $M_j \in M$ models a sensing state (i.e., mode) in which mobile devices behave at a particular point in time. Σ set represents some sensor input readings, a notice of a context switching to another state, or an occurrence of an anticipated event such as getting closer to one location of a business. $\delta : M \times \Sigma \rightarrow M$ is a function that represents a transition from the current state to another one. Σ set represents the newly sensed data that fires the triggers for a state change. Figure 2 illustrates an example of an FSM that has three modes. As shown in the figure, the sensing process of a particular data could fire a trigger $t_i \in \Sigma$ that causes a transition from mode M_i to mode M_j.

The business owner side is implemented as a web application, which uses GeoJSON for modeling the businesses' geolocations. Each GeoJSON object represents some features and attributes of a business, which includes the location coordinates, location size, and location boundaries. Each business is defined by coordinates that shape polygon borders on top of the Google Map.

Business owners can precisely pick out the location coordinates of each business on the map using the developed GUI at the business side. A cloud-based database is used to store these coordinates. When a new business is initiated, the selected coordinates of the drawn polygon are sent to the user side (i.e., mobile app) as a GeoJSON file communicated through the developed REST service.

We developed a python script at the mobile app to parse the incoming GeoJSON files to generate the place polygon coordinates. Then, these coordinates are sent to the mobile app, which uses Google Maps API for displaying the business location on Google Map. An example of one GeoJSON file that is used in our system is shown in Fig. 3. As shown in the figure, each GeoJSON object represents different spatial attributes of the bounded entity of a business location.

3 Implementation

We implemented the business side as a web-based app, which provides an interactive dashboard for businesses to initiate a fresh advertisement, edit an advertisement, display responses and inquiries to their advertisements, and response to inquiries

```
geo_json = {"type": "Feature", "geometry": {"type":
"Polygon", "coordinates":[(30.095319, -95.993581),
(30.096340, -95.993399), (30.096015, -95.991854),
(30.095040, -95.992219), (30.095319, -95.993581)]},
        "properties": {"name": " Pizza Restaurant No 1",
          "styleUrl": "#poly-4F2682-3000-128",
          "styleHash": "-50cd947a",
          "styleMapHash": {
             "normal": "#poly-4F2682-3000-128-normal",
             "highlight": "#poly-4F2682-3000-128-highlight"
          },
          "description": "Pizza Restaurant location",
          "stroke": "#4f2682",
          "stroke-opacity": 1,
          "stroke-width": 3,
          "fill": "#4f2682",
          "fill-opacity": 0.5019607843137255
    }
}
```


Fig. 3 An Example of a GeoJSON file

and requests. We used HTML5, PHP 7.4, CSS3 and JS, MySQL 8.0, in addition to Bootstrap [16] and JSON to develop the business side.

Figure 4a illustrates the registration form for business owners. The forms ask business owners their name, email address, telephone #, and username and password.

Figure 4b illustrates the form that is used to initiate an advertisement on the business side. This form allows businesses to draw the place's polygon using the interactive map for specifying the boundary for their business locations. Business owners need to enter: (1) the name of their businesses that would appear on the mobile side; (2) the type of the created ad such as garage sale, restaurant opening, and job; (3) other attributes, such as description, business photos, time frame, etc.

Figure 4c illustrates the login page at the cloud side for business owners. Similarly, we implemented a login page that utilized the `$_POST` technique to authenticate users. Figure 4d illustrates the existing advertisements list created by a business owner. Owners are also allowed to see the current status of their advertisements and edit and delete the existing advertisements. This page illustrates the title, description, type, time frame, price, address, and geographical business location.

We used the Android ADT bundle Development Environment (64 bit) to implement the mobile app on the user side. The mobile application utilizes various technological tools and programming languages: (1) Android SDK to develop the front-end mobile user activities; (2) PHP 7.4 for implementing the middle-ware between the database server and the mobile app; and (3) MYSQL 8.0 to implement the system database.

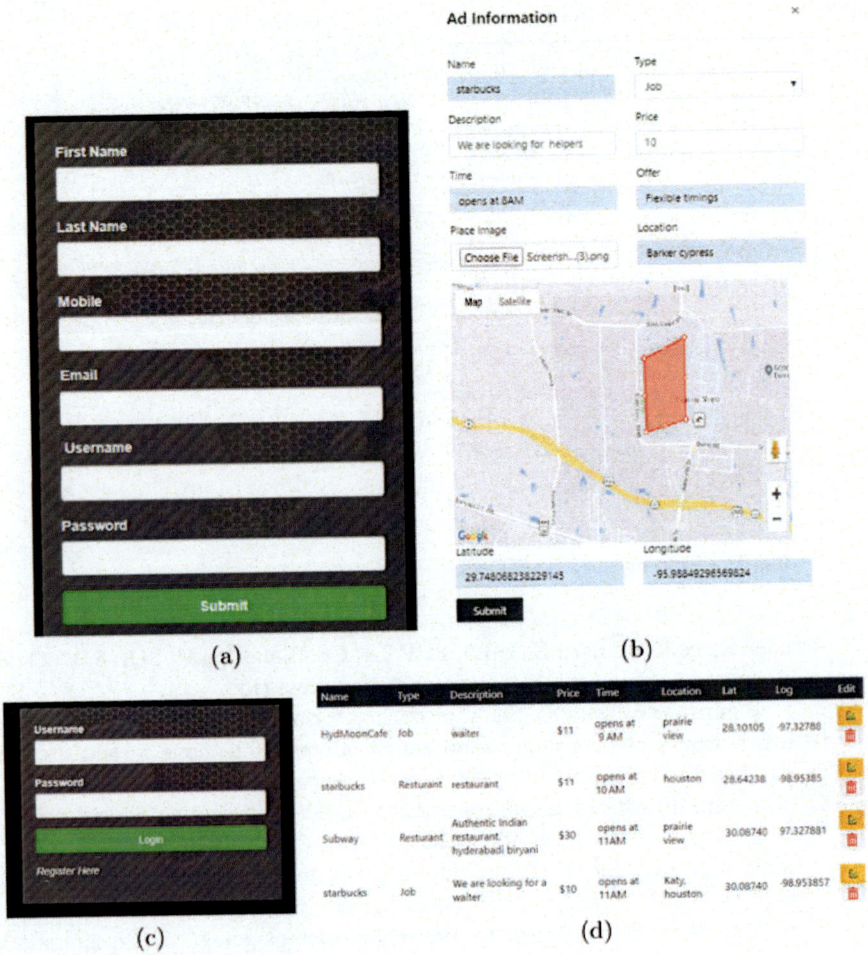

Fig. 4 The implementation of the cloud-based side. (**a**) Registration. (**b**) Creating a new Ad. (**c**) Login. (**d**) List of Ads

Figure 5a illustrates the mobile app landing form. The form has a search `textarea`, which allows mobile users to search for and navigate to any destination provided by the Google Maps API. As shown in Fig. 5b, the autocomplete option is used in the search functionality to increase the mobile user experience using our LBA system.

Users can commute to their target destination by tapping a button titled `START JOURNEY`. The mobile application then creates the path between the mobile's current geolocation and the destination. This process is carried out; meanwhile, the path linking between the start and destination points is being rendered. An example

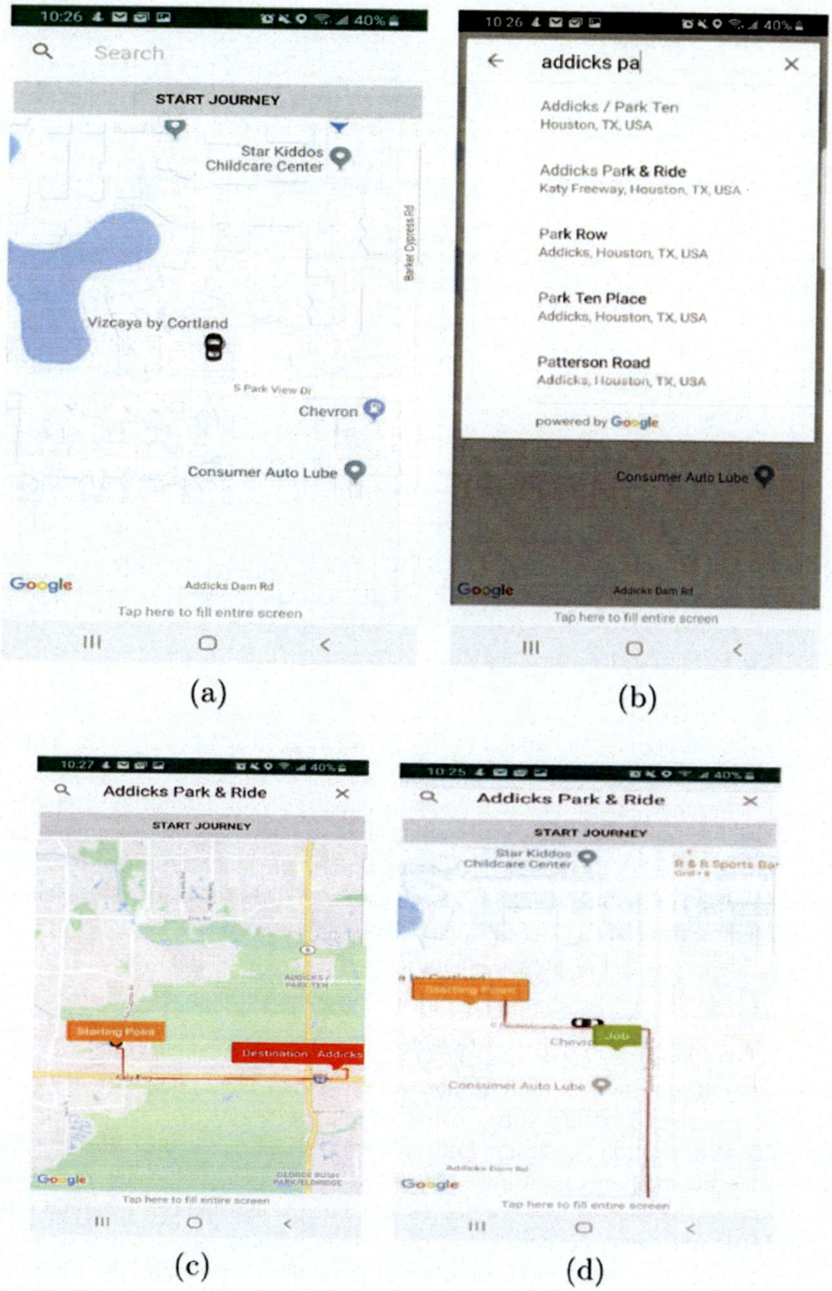

Fig. 5 Screenshots of the mobile user application. (**a**) Landing page. (**b**) Selecting a destination place. (**c**) Displaying the route map. (**d**) Showing an advertisement during the driving state

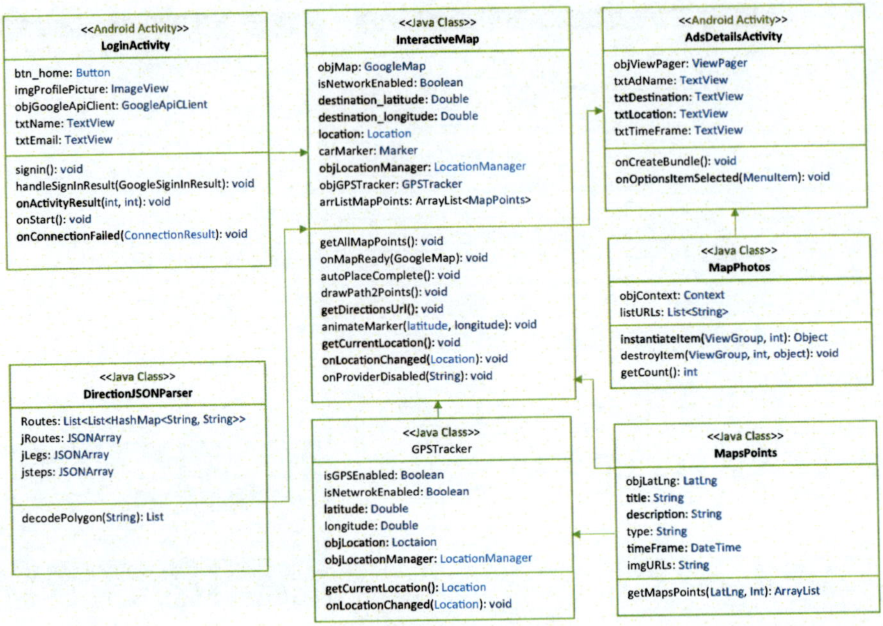

Fig. 6 The class diagram of the mobile side

of this process is illustrated in Fig. 5c where `Addicks / Park Ten` is selected as a target destination.

A screenshot of the mobile app displaying an ad in the driving mode is shown in Fig. 5d. Figure 5d shows different advertisement categories rendered on the app's map. When a mobile user taps on a particular advertisement, the app displays another form that contains more details of this selected advertisement such as photos, time frame, and detailed description of the ad.

Figure 6 shows the class diagram of the system on the mobile side. Two types of class entities are defined in our LBA system, namely Android Activities and Java Classes.

As illustrated in Fig. 6, the `LoginActivity` enables mobile users to login into the system via their Google accounts. Then, they are redirected to the `InteractiveMap` form, using which users will be able to search for a particular location or destination. The `InteractiveMap` class is used to render the path between the current user location and the target location. The `Advertisements- DetailsActivity` represents the mobile form that displays the advertisement information (e.g., photos and detailed description). The `DirectionJSONParser` class is developed for parsing the geolocation of the landmarks and the target location, which Google Maps support. We created the `Routes` list to load the resulting places that were generated by the map. Each item in Routes contains an array of polygon coordinates that states the boundaries of that place location on Google Maps.

We developed the GPSTracker class to get the current geolocation of mobile users (i.e., longitude and latitude). This information is stored in the LatLng object temporarily before sending it to the database server on the cloud. We developed the MapPhotos class to render the advertisement photos on Google Maps. First, the photo URLs are fetched from the database and then sent to the mobile app to display them when the user clicks on an ad. We implemented both MapsPoints and InteractiveMap classes to be used in rendering the different advertisements on the mobile map. We used JSON to serialize all the details of the selected advertisement, which are communicated between the server and the mobile app. Once the JSON file is received at the mobile app, it will be converted into Java objects using the Retrofit API [36].

4 Evaluation

We conducted a set of experiments to evaluate the performance and scalability of our LBA system. We installed some instrumentations inside the mobile app in order to calculate the CPU processing time that is taken to execute the different processing jobs, including the sensing data collection, context-switching process for the multi-modal approach, acquiring the geolocation coordinates, and making the mobile user trajectories anonymous. Also, a set of instrumentations were developed in the app to calculate the power consumption used by these sensors: gyroscope, GPS, and accelerometer. We carried out every experiment presented in this section for 10 trials. Then, we took and showed the average of the trials' output.

We used a case study for the experimental evaluation. It involves several human activities to mimic various states for the mobile user when viewing advertisements on the map. The mobile application is executed on top of a smartphone running Android 10, which has the following specs: Samsung S8, 4 GB RAM, LTE Category 16, Adreno 540 GPU, and 2.3 GHz Octa-core CPU. We defined 4 activity modes, namely driving, bicycling, walking, and stilling.

Every state (mode) has different requirements in terms of sensors (e.g., accelerometers, gyroscopes, and GPS sensors). The stilling mode requires only the accelerometer data. The walking mode requires both the gyroscope and accelerometer and data. Both the driving and bicycling states require both GPS and accelerometer data.

Additionally, the logic of state transition is mainly based on sensor readings from the three sensors. It is assumed that the three sensors are sampled every one second at a sampling rate of 1 Hz. In other words, the sensors that are needed for the current state (e.g., driving) functionality are sampled at a sampling rate of 1 Hz to detect any triggers for mode transition. For the other sensors, it is required to collect fresh sensor data as described in [29].

4.1 Performance

In order to estimate the scalability of our LBA system on the mobile side, we calculated different processing resources required to create a single ad. Separately, we calculated the ongoing monitoring cost to detect any triggers for state transition. In addition, we measured the processing CPU time and the sensing overheads required to execute the instructions for triggering the state transitions.

We measured processing time for the ongoing process of monitoring the state transition triggers. These processing times are shown in Table 1.

The CPU time (measured in milliseconds) is taken to acquire, process, and examine the three sensor readings to detect any triggers for mode transition. The tag *Old* happens in the case of utilizing the already collected sensor readings for the current state functionality. The tag *New* occurs in the case of collecting new sensor data to detect mode transition triggers only. Both the sampling rate of the mode functionality and evolution was set to be 1 Hz. The time taken for transitioning each sensor from the existing sampling rate to a different rate was calculated to be 6.21 ms (with a standard deviation of 1.21) for the accelerometer sensor, 10.71 ms (with a standard deviation of 1.76) for the gyroscope sensor, and 17.39 ms (with a standard deviation of 2.36) for the GPS sensor.

The overhead costs of making a transition from one mode to another one are displayed in Table 2. These overhead costs include the changes in the sampling rate of the involved sensors and the processing time needed to trigger that change. We also measured the processing time for the change in the sampling rate of all involved sensors. This happens only if the sensors used in the mode functionality cannot be reused for the mode transition to a new mode. Note that the processing time between any pair of states is symmetric.

To put these overhead costs in some perspective, around 28 states (modes) per every second, on average, can be hosted on a smartphone of our modest configuration. For instance, when a mobile app needs around 10 sensor samples every second, which could be collected from different sensors, our LBA system could support about 2.8 applications simultaneously. In the case of 1 sensor sample

Table 1 The processing time for the ongoing process of monitoring the state transition triggers (in *milliseconds*)

Mode	Accelerometer	Gyroscope	GPS	Total
Stilling	Old: 0.39	New: 4.50	New: 7.31	12.2
Walking	Old: 0.39	Old: 0.39	New: 7.31	8.09
Bicycling	Old: 0.39	New: 4.50	Old: 0.39	5.28
Driving	Old: 0.39	New: 4.50	Old: 0.39	5.28

Table 2 The overhead costs of making a transition from one mode to another one (measured in *milliseconds*)

	Stilling	Walking	Bicycling	Driving
Stilling	X	12.32	19	19
Walking	12.32	X	31.32	31.32
Bicycling	19	31.32	X	1.61
Driving	19	31.32	1.61	X

Table 3 The overhead energy consumption of our mobile app (measured in *milli-joule*)

	Accelerometer	Gyroscope	GPS
Cost per transition	2.42	3.18	4.05
On-going cost per feed set	0.63	1.24	1.93

(i.e., 1 hz), our LBA system could support around 115 applications simultaneously on one smartphone.

4.2 *Power Consumption Overhead*

The overhead energy consumption of our mobile app is displayed in Table 3. It were calculated to be 0.63 mJ for the accelerometer sensor, 1.24 mJ for the gyroscope sensor, and 1.93 mJ for the GPS sensor. We found that the overhead costs of switching the sampling rate of the accelerometer sensor to be 2.42 mJ, the gyroscope sensor to be 3.18 mJ, and the GPS sensor to be 4.05 mJ.

4.3 *Overhead Analysis*

Without the actual sensing, we calculated the power overheads caused by our mobile app for trigging state transitions. We used this overhead analysis to measure the energy non-sensing overheads of the proposed multi-modal context-switching approach.

The average power used by the multi-modal approach were measured to 70.4 mJ and 79.6 mJ for the accelerometer and gyroscope sensors, respectively. This was approximately 4% of the total power demand of the accelerometer sensor experiments and 0.8% for the power demand of the gyroscope sensor experiments. The significant power difference is justified by the order-of-magnitude more massive power overheads of the gyroscope sensor.

5 Conclusions

In this chapter, we presented an LBA-based system that would help businesses advertise their projects and services affordably. We implemented two types of applications: a web-based application running on a server hosted in the cloud and a mobile-based app running on smartphones. Also, we presented the multi-modal sensing approach to represent and program mobile apps based on sensor context switching. The main objective of the proposed multi-modal sensing approach is to separate the sensing and functionality concerns, which makes developing a context-switching app more modular.

Several sets of experiments were conducted to assess the scalability and performance of our LBA system at the mobile user side. The experiments' results illustrated that the developed LBA system is scalable and highly responsive in hosting numerous advertisements in the system.

For future work, we are looking at ways of how we can compose ModeSens [30], a multi-modal sensing approach, with ShareSens [27] in order to provide the ability to share sensor data between a large number of apps running on the same mobile device. The combination of ShareSens and ModeSens would be beneficial for backing the sensing requirements of a broad domain of research work [3, 6, 8, 23–25, 28, 32] and applications [1, 2, 5, 7, 9, 11–13, 26, 31]. Finally, extensive experiments will be conducted using big datasets to further study the robustness of our LBA system.

Data Availability

The data and the source code are available online for the public use at: https://github.com/ahmed-pvamu/Location-based-Mobile-Advertising-System

Acknowledgments National Science Foundation (NSF) partially supports this research work under award number 2011330.

References

1. A. Abdelmoamen, A modular approach to programming multi-modal sensing applications, in *Proceedings of the IEEE International Conference on Cognitive Computing San Francisco* (2018), pp. 91–98
2. A. Abdelmoamen, N. Jamali, A model for representing mobile distributed sensing-based services, in *Proceedings of the IEEE International Conference on Services Computing, San Francisco* (2018), pp. 282–286
3. A. Abdelmoamen, D. Wang, N. Jamali, Approaching actor-level resource control for Akka, in *Proceedings of the IEEE Workshop on Job Scheduling Strategies for Parallel Processing, Vancouver* (2018), pp. 1–15
4. AdWords: grow business with Google ads https://ads.google.com/home/. Accessed 26 Oct 2021
5. A.A. Ahmed, A model and middleware for composable IoT services, in *Proceedings of the International Conference on Internet Computing & IoT, Las Vegas* (2019), pp. 108–114
6. A.A. Ahmed, A privacy-preserving mobile location-based advertising system for small businesses. Eng. Rep. **e12416**, 1–15 (2021). https://doi.org/10.1002/eng2.12416
7. A.A. Ahmed, G. Agunsoye, A real-time network traffic classifier for online applications using machine learning. Algorithms **14**(8) (2021). https://doi.org/10.3390/a14080250
8. A.A. Ahmed, M. Echi, Hawk-eye: an AI-powered threat detector for intelligent surveillance cameras. IEEE Access **9**, 63283–63293 (2021). https://doi.org/10.1109/ACCESS.2021.3074319

9. A.A. Ahmed, T. Eze, An actor-based runtime environment for heterogeneous distributed computing, in *Proceedings of the International Conference on Parallel & Distributed Processing, Las Vegas* (2019), pp. 37–43
10. A.A. Ahmed, N. Jamali, CSSWare: a middleware for scalable mobile crowd-sourced services, in *Proceedings of the 7th EAI International Conference on Mobile Computing, Applications and Services (MobiCASE'15), Berlin* (2015), pp. 181–199
11. A.A. Ahmed, G.H. Reddy, A mobile-based system for detecting plant leaf diseases using deep learning. AgriEngineering **3**(3), 478–493 (2021). https://doi.org/10.3390/agriengineering3030032
12. A.A. Ahmed, A. Olumide, A. Akinwa, M. Chouikha, Constructing 3d maps for dynamic environments using autonomous UAVs, in *Proceedings of the 2019 EAI International Conference on Mobile and Ubiquitous Systems: Computing, Networking and Services (MobiQuitous'19), Houston* (2019), pp. 504–513
13. A.A. Ahmed, S.A. Omari, R. Awal, A. Fares, M. Chouikha, A distributed system for supporting smart irrigation using IoT technology. Eng. Rep. **3**, 1–13 (2020). https://doi.org/10.1002/eng2.12352
14. G. Aydin, B. Karamehmet, A comparative study on attitudes towards SMS advertising and mobile application advertising. Int. J. Mobile Commun. **15**(5), 514–536 (2017)
15. L. Besaleva, A. Weaver, CrowdHelp: a crowdsourcing application for improving disaster management, in *Proceedings of the IEEE Conference on Global Humanitarian Technology* (2013), pp. 185–190
16. Bootstrap: an open-source framework for designing front-end web application https://getbootstrap.com/. Accessed 26 Oct 2021
17. A.D. Carli, M. Franco, A. Gassmann, C. Killer, B. Rodrigues, E. Scheid, D. Schoenbaechler, B. Stiller, WeTrace – a privacy-preserving mobile covid-19 tracing approach and application (2020). arXiv:2004.08812
18. S. Dhar, U. Varshney, Challenges and business models for mobile location-based services and advertising. Commun. ACM Mag. **54**(5), 121–128 (2011)
19. W. Gong, B. Zhang, C. Li, Location-based online task assignment and path planning for mobile crowdsensing. IEEE Trans. Veh. Technol. **68**(2), 1772–1783 (2019)
20. R.Y. Li, S. Liang, D.W. Lee, Y.J. Byon, TrafficPulse: a mobile GISystem for transportation, in *Proceedings of the ACM SIGSPATIAL International Workshop on Mobile Geographic Information Systems* (2012), pp. 9–16
21. S. Meng, W. Huang, X. Yin, M.R. Khosravi, Q. Li, S. Wan, L. Qi, Security-aware dynamic scheduling for real-time optimization in cloud-based industrial applications. IEEE Trans. Ind. Inf. **17**(6), 19–28 (2021)
22. M. Min, R. Sasank, S. Katie, Y. Nathan, B. Jeff, E. Deborah, H. Mark, H. Eric, W. Ruth, P. Boda, PEIR: the personal environmental impact report, as a platform for participatory sensing systems research, in *Proceedings of the ACM Conference on Mobile Systems, Applications, and Services MobiSys, Poland* (2009)
23. A.M.A. Moamen, H.S. Hamza, On securing atomic operations in multicast AODV. Ad-Hoc Sensor Wireless Netw. **28**, 137–159 (2015)
24. A.A. Moamen, N. Jamali, An actor-based approach to coordinating crowd-sourced services. Int. J. Serv. Comput. **2**(3), 43–55 (2014)
25. A.A. Moamen, N. Jamali, CSSWare: a middleware for scalable mobile crowd-sourced services, in *Proceedings of MobiCASE, Berlin* (2015), pp. 181–199
26. A.A. Moamen, N. Jamali, CSSWare: an actor-based middleware for mobile crowd-sourced services, in *Proceedings of the 2015 EAI International Conference on Mobile and Ubiquitous Systems: Computing, Networking and Services (Mobiquitous'15), Coimbra* (2015), pp. 287–288
27. A.A. Moamen, N. Jamali, ShareSens: an approach to optimizing energy consumption of continuous mobile sensing workloads, in *Proceedings of the 2015 IEEE International Conference on Mobile Services (MS'15), New York* (2015), pp. 89–96

28. A.A. Moamen, N. Jamali, Supporting resource bounded multitenancy in Akka, in *Proceedings of the ACM SIGPLAN International Conference on Systems, Programming, Languages and Applications: Software for Humanity (SPLASH Companion 2016)* (2016), pp. 33–34
29. A.A. Moamen, N. Jamali, Opportunistic sharing of continuous mobile sensing data for energy and power conservation. IEEE Trans. Serv. Comput. **13**, 503–514 (2020)
30. A.A. Moamen, J. Nadeem, ModeSens: an approach for multi-modal mobile sensing, in *Companion Proceedings of the 2015 ACM SIGPLAN International Conference on Systems, Programming, Languages and Applications: Software for Humanity* (ACM, Pittsburgh, 2015), pp. 40–41
31. A.A. Moamen, H.S. Hamza, I.A. Saroit, Secure multicast routing protocols in mobile ad-hoc networks. Int. J. Commun. Syst. **27**(11), 2808–2831 (2014)
32. A.A. Moamen, D. Wang, N. Jamali, Supporting resource control for actor systems in Akka, in *Proceedings of the International Conference on Distributed Computing Systems (ICDCS 2017)* (2017), pp. 1–4
33. MobileAds: third-party ad serving platform for rich media and video advertising https://www.mobileads.com/. Accessed 26 Oct 2021
34. D. Paulino, A. Reis, J. Barroso, H. Paredes, Mobile devices to monitor physical activity and health data, in *Proceedings of the 12th Iberian Conference on Information Systems and Technologies (CISTI)* (2017), pp. 1–4
35. PokemonGo: an augmented reality mobile game https://www.pokemongo.com/en-us/. Accessed 26 Oct 2021
36. Retrofit: a type-safe http client for Android and Java https://square.github.io/retrofit/. Accessed 26 Oct 2021
37. Waze: a GPS navigation software app owned by Google https://www.waze.com/. Accessed 26 Oct 2021
38. K. Xu, W. Zhang, Z. Yan, A privacy-preserving mobile application recommender system based on trust evaluation. J. Comput. Sci. **26**, 87–107 (2018)
39. M. Zook, M. Graham, T. Shelton, S. Gorman, Volunteered geographic information and crowdsourcing disaster relief: a case study of the Haitian earthquake. J. World Med. Health Policy **2**(2), 7–33 (2010)

Cyber Threat Intelligence Methodologies: Hunting Cyber Threats with Threat Intelligence Platforms and Deception Techniques

Arturo E. Torres, Francisco Torres, and Arturo Torres Budgud

Abstract Faced with the great wave of cyber threats, as well as the considerable increase in cybercrime in recent years, organizations have been forced to redefine their digital defense strategies to protect their information assets, infrastructure, and reputation from different people—malicious adversaries. Given this, the IT cybersecurity community has chosen to use intelligence techniques to prepare for emerging cyber threats. Therefore, the field of Cyber Threat Intelligence (CTI) has had significant growth in recent years, given the growth and evolution of cyber threats, as well as the complexity of the techniques used by adversaries. However, the CTI field has different challenges for companies that don't have a big budget or lack the experience to implement a CTI plan. The main contribution of this research is based on the compilation and investigation of the schemes, tools, challenges, and sets of methodologies most used for the execution of a CTI program, as well as the deployment of a CTI platform based on deception techniques (honeypots) for data collection and cyber threat events. This enables organizations with smaller budgets to use the CTI platform and the methodologies described in this document to stay secure.

Keywords Cyber Threat Intelligence (CTI) · Information technology · Cyber security events · Cyber threats · Intelligence sources · Cyber security · Deception techniques · Honeypots · Etc

---

A. E. Torres (✉) · F. Torres · A. T. Budgud
Universidad Autonoma de Nuevo Leon, San Nicolás de los Garza, N.L., Mexico
e-mail: arturo.torrescv@uanl.edu.mx; francisco.torresgrr@uanl.edu.mx; arturo.torresbg@uanl.edu.mx

F. Torres-Guerrero et al. (eds.), *2nd EAI International Conference on Smart Technology*, EAI/Springer Innovations in Communication and Computing,
https://doi.org/10.1007/978-3-031-07670-1_2

1 Introduction

In recent years, organizations have made significant progress in many aspects at a global level, thanks to the accelerated growth and constant use of technology in their day-to-day life, that is, digital tools have become critical for organizations and your daily operation. For example, in 2019, more than 3.9 billion email accounts were registered, with an approximate of 559,000 emails sent and received every second during that year, and it is expected that by 2024 the figure will reach 4.3 billion email accounts [1]. However, dependence on digital information and tools brings with it a large number of challenges that organizations must face, since, in recent years, trends in digital crimes and cyber threats have increased considerably. Following the example mentioned above of the use of email, according to Internet Live Stats [2], approximately 2,962,154 emails were sent per second in 2020, with 67% of these being classified as spam. This tells us that digital systems are being pressured by adversaries and not only in quantity, but also in the complexity of the techniques used, such as phishing or Business Email Compromise (BEC), to be able to compromise the systems or steal corporate information [3], this being the main vector of digital fraud and malware distribution in recent years with a recorded loss of $ 1.25 trillion in 2018 alone.

Unfortunately, dependence on technology, as well as the large amount of critical and confidential information handled by organizations today, has become the target of cybercriminals, who have developed new ways to affect integrity, availability, and confidentiality of the organizations' systems and data, using advanced techniques such as digital hijacking, better known as ransomware, which recorded an approximate loss of 10.1 billion euros in 2019 [4]. The main problem that organizations face is that advanced cyber threats bypass the protection controls implemented in organizations, such as firewalls and antivirus, which are commonly based on detection engines for static signatures or known threats. Therefore, cybersecurity specialists have chosen to develop new ways to prepare for and generate new strategies in the face of these advanced threats, using collaborative intelligence platforms to prevent and/or minimize the risk of a cyberattack before it happens. This practice is known as Cyber Threat Intelligence (CTI), which is defined as knowledge based on evidence, which includes context, mechanisms, indicators, implications, and practical advice, about an existing or emerging threat to information assets of organizations that can be used to inform decisions regarding the subject's response to that threat [5].

1.1 Contributions

The main contribution of this research is based on the compilation and investigation of the schemes, tools, challenges, and sets of methodologies most used for the execution of a CTI program. This will allow organizations with smaller budgets to use the CTI platform and the methodologies described in this document to generate a defense strategy and stay secure. This chapter is organized as follows:

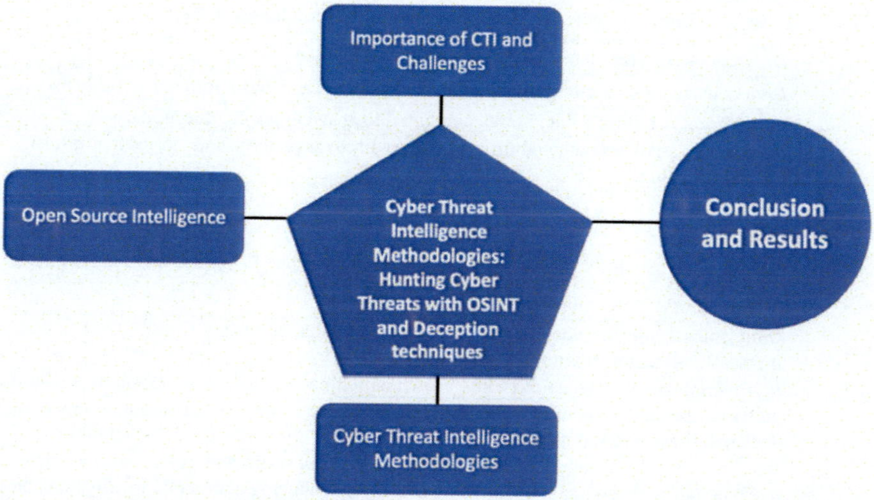

Fig. 1 Contributions and research structure

- Section II raises the importance of the CTI field, as well as the challenges that generating intelligence represents for organizations.
- Section III aims to present the research of the most relevant methodologies in the field of CTI during the last years.
- Section IV's contribution is the deployment of a TIP exposed to the internet based on deception techniques for the collection of cyber threat events for the generation of CTI.
- Section V presents the results obtained from the investigation.
- Section VI presents the conclusions of the research article (Fig. 1).

2 Importance and Challenges of CTI

The CTI field has been widely accepted by different cybersecurity specialists from different industry sectors, such as the energy sector, which has been subject to different types of advanced cyberattacks, for which reason research has been carried out to develop CTI platforms that integrate the strategic, tactical, and operational levels of IT, aiming to provide a comprehensive response to the evolving threat landscape of energy systems [6]. On the other hand, Verizon confirmed 927 incidents of cyberattacks and about 207 cases of disclosure of confirmed unauthenticated data in the financial sector, for which various cyber defense strategies and techniques for the financial sector have been explored [7], as well as the integration of digital systems, Internet of Things (IoT), cloud computing paradigms to develop smart systems, smart homes and smart cities [8]. However, one of the main points that needs to be

Table 1 Differences between data, information, and intelligence [9]

Differences between data, information, and intelligence
Data consists of discrete facts and statistics collected as a basis for further analysis.
Information is multiple data points combined to answer specific questions.
Intelligence analyzes data and information to discover patterns and stories that inform decision-making.

Table 2 Cybersecurity: Data, information and intelligence [9]

Cybersecurity: Data, information and intelligence
Data is usually just indicators like IP addresses, URLs, or hashes. It doesn't tell us much without analysis.
Information answers questions like, "How many times has my organization been mentioned on social media this month?" Although this is a much more useful result than raw data, it still does not directly report a specific action.
Intelligence is the product of a cycle of identifying questions and objectives, gathering relevant data, processing and analyzing that data, producing actionable intelligence, and distributing that intelligence.

understood in this area is the difference between data, information, and intelligence to understand CTI. Therefore, we can define data as an individual element that contains information on either a system, an action or an executed process, that is, individual elements with a specific meaning. On the other hand, we can define the term threat as the possible danger that can be used to exploit an existing vulnerability with the intention of causing damage to systems, networks, or entire organizations (Table 1).

These three terms are sometimes used without much attention according to Dr. Christopher Ahl-berg [9], which explains that some threat feeds are advertised as intelligence when in reality they are just data packets. Organizations often embed threat data sources into their network only to find that they cannot process all the additional data, which only adds to the burden on analysts trying to classify threats. Rather, threat intelligence lightens that burden by helping analysts decide what to prioritize and what to ignore. Therefore, a different context can be given to the terms data, information, and intelligence when talking about cybersecurity (Table 2).

Therefore, we can say that CTI is the ability to acquire knowledge about a company, as well as its existing conditions and capabilities, in order to determine the possible actions of a malicious actor or threat when exploiting existing critical vulnerabilities. In addition, it uses multiple information security disciplines (threat intelligence, vulnerability management, security configuration management, incident response, etc.) and sets of tools to collect information about the network through monitoring and reporting to provide decision makers at all levels to prioritize resource allocation to perform risk mitigation.

2.1 CTI Challenges

The main challenge in executing a CTI strategy focuses on the quality of intelligence obtained through data analysis and lies mainly in being able to transform this large amount of information into something actionable that can be used to make decisions for top management [10], for example, being able to prioritize activities and assign budget or personnel based on the impacts that may be had based on the data collected. This requires not only time and/or effort from the IT or cybersecurity team, but also organization, collaboration between the different areas, and experience and resources assigned by senior management to carry out these investigations successfully. Although, the commercialization of products and services related to CTI from different developers and manufacturers has helped automate many of the tasks related to the extraction, detection, and update of threats, and especially the automation of responses to incidents [11].

In order to significantly interrupt or prevent the attack or intrusion of the adversary, it is necessary to have a defensive strategy and prepare the infrastructure to take into account the security needs, controls, processes, and resources that must be available. With knowledge of the adversary's tactics and objectives, defenders must prepare their infrastructure to counter attacks in the widest range possible to cover all possible adversaries' attack vectors. However, there is some research that mentions this as a great challenge, since it is mentioned that the CTI field lacks a mature methodology, which can affect the analysis of threats and adversaries by defenders [12]. Despite its challenges, CTI should not be ruled out yet, since it is an emerging field that has great potential and constant development by the cybersecurity analyst community, with the objective of applying defense strategies and controls against adversaries. seeking to engage an organization

3 CTI Methodologies

There are different methodologies in the field of CTI, which could be defined as a structure to think about how attackers operate, discover their methods and in which part of the general life cycle of the attack that event is occurring. More specifically, CTI methodologies allow us to be very prescriptive in how we attack a specific situation, that is, they allow us to focus attention on the appropriate areas to ensure monitoring and mitigation of existing or emerging threats. They also provide a common language to communicate internally and also externally regarding threat details, interrelationships between events, and correlations with external data sources. So, we can say that they allow us to connect and understand where something is happening and focus our resources within that small area rather than trying to take a reactive approach. In addition, it allows us to be much more focused on the specific area that needs our attention to assign specific resources to a threat or technique used by an adversary that could harm our organization. That way, defenders don't waste time, effort, and resources working in areas that are not necessarily affected or perhaps not necessarily relevant to the incident they are facing or are about to face.

3.1 Cyber Kill Chain Model

The methodology developed by Lockheed Martin [13] is based on the military concept of "Kill Chain," which consists of seven different areas that allow us to understand in which part of the process or attack chain a specific threat occurs, whether in reconnaissance, weaponry, delivery, etc. Therefore, if we understand where that threat is in the process, we can focus our resources and our efforts on mitigating it. And if we have a proper framework, we can understand what actions need to be taken in that area so that we can respond quickly to the opponent's techniques [14]. For example, first comes recognition, where the adversary is looking for a weakness, that is, obtaining registration credentials or information that can be used for a phishing attack, where in this case, the weakness is the user who you receive the phishing added to the vulnerabilities of your device. The next thing is weaponization, which consists of creating the delivery of the threat, using an exploitation technique (exploit), such as a Backdoor typically. Delivery is the process of sending that payload to the victim, which could be a malicious email, it could be a USB memory left on a desk on the floor of a parking lot near the organization or even in the parking of this. Then we have vulnerabilities, in which the attacker performs the act of executing the code on the remote system and then proceeds to the phase of the actual installation of malware in that target. And that brings us to C&C, which will create a channel or persistence where the attacker can control the system remotely by sending instructions with some specific objective or purpose. At that point, the attacker already has control of perhaps more than one system; therefore, it is very important to monitor the different C&C channels detected in our network. And at the end of this, the desired actions are performed. So that's the intended goal, whether it's encrypting data, destroying it, exfiltering it, etc (Fig. 2).

This methodology describes an intelligence-based, threat-focused approach to studying intrusions from the perspective of adversaries. Each phase of the intrusion is assigned to courses of action for detection, mitigation, and response and shows that there are certain phases that the adversary has to fulfill in order to complete its objective, and that in addition, they have been used in different fields of the industry to generate strategies based on taxonomies of specific threats such as Trojans that affect the financial sector [16] and various investigations are explored on how to be able to predict certain threats or adversaries' behavior [17], which leads us to think that this model is based on the reconstruction of an attack in order to better understand it and mitigate it. Therefore, if we can understand where a specific action is in the process, we will know how to focus our efforts and resources to mitigate

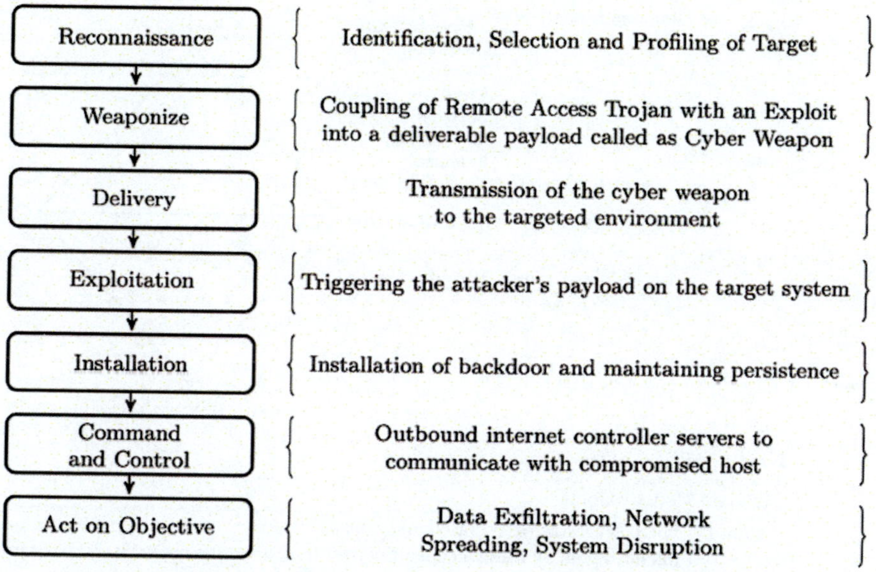

Fig. 2 Cyber Kill Chain Model phases [14, 15]

the threat before it reaches a final phase, such as a link in a chain is broken. the objective is to break the progress of the cyberattack in any of its phases [18].

3.2 *The Diamond Model of Intrusion Analysis*

Intrusion analysis has long been considered an art to be learned and practiced, rather than a science to be studied and refined. However, approaching it only as an art has long-delayed improvements and understanding, further slowing down the evolution of threat mitigation that relies on efficient, effective, and accurate analysis. Unknowingly, analysts have used the Diamante model for decades, but have lacked the full framework to understand, improve, and focus their efforts. This model describes the main capacities and characteristics of an intrusion event: adversary, capacity (techniques and tools used by an adversary), infrastructure, and victim, which are linked in a diamond-shaped diagram, in which the edges are used to represent relationships between features that can be exploited analytically to discover and further develop awareness of malicious activity [19], that is, the model describes that an adversary deploys a capacity on some infrastructure against a victim (Fig. 3).

These activities, in turn, are known as events, which define a series of steps that the adversary must execute to achieve their objective. Likewise, the authors of the model describe 7 fundamental bases to understand the intrusion model process in

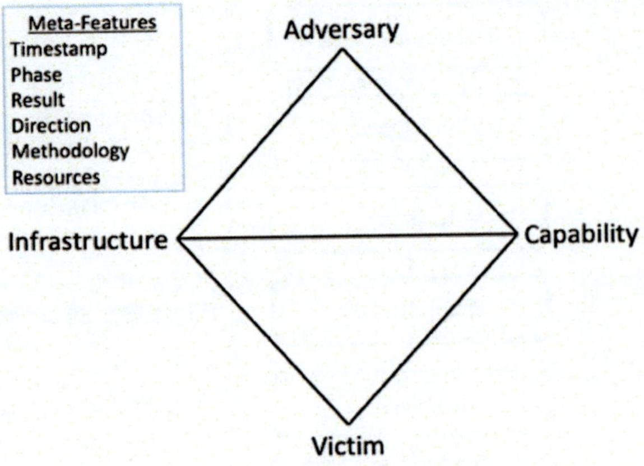

Fig. 3 The Diamond Model of intrusion analysis [19]

Table 3 Axioms of the Diamond Model [19]

Diamond Model Axioms	
Axiom 1	For every intrusion event, there exists an adversary taking a step towards an intended goal by using a capability over infrastructure against a victim to produce a result.
Axiom 2	There exists a set of adversaries (insiders, outsiders, individuals, groups, and organizations) that seek to compromise computer systems or networks to further their intent and satisfy their needs.
Axiom 3	Every system, and by extension every victim asset, has vulnerabilities and exposures.
Axiom 4	Every malicious activity contains two or more phases that must be successfully executed in succession to achieve the desired result.
Axiom 5	Every intrusion event requires one or more external resources to be satisfied prior to success.
Axiom 6	A relationship always exists between the adversary and their victim(s), even if distant, fleeting, or indirect.
Axiom 7	There exists a subset of the set of adversaries that have the motivation, resources, and capabilities to sustain malicious effects for a significant length of time against one or more victims while resisting mitigation efforts. Adversary-victim relationships in this subset are called persistent adversary relationships.

the form of Axioms, where the objectives and/or needs of the adversaries are defined in order to meet their objectives (Table 3).

The model establishes a formal method that applies scientific principles to the analysis of threats, in particular, those of measurement, testability, and repeatability, providing a comprehensive method of documentation, synthesis, and correlation of the activity [20]. For this reason, we can say that both the events and the processes or threads of each activity carried out by the attacker are necessary elements for a complete understanding of the malicious activity itself, since a more effective and strategic mitigation requires an understanding and context of the intrusions themselves, with the main objective of being able to expand the panorama of the threat and understand the phases and processes of this. This scientific approach and simplicity produce improvements in analytical effectiveness, efficiency, and precision. Ultimately, the model provides opportunities to integrate and generate real-time intelligence for network defense, automating correlation between events, confidently classifying events in adverse campaigns, and forecasting adverse operations while planning and playing mitigation strategies.

While the Kill Chain model provides information on the attackers' operations, the Diamante model broadens the perspective and context of the attackers between each of the intrusion phases, that is, it together allows to have a broader view of the attacker and not just the technical indicators. In addition, the Diamante model provides a formal mathematical method for the analysis and grouping of effective graphs (e.g., grouping/ranking) to solve many kinds of analytical problems. [19]. The Diamond Model identifies how and why an attack occurs, since we can see that an attacker attacks a victim based on two main attributes, called infrastructure and capacity, precisely capturing and organizing the fundamental concepts that underpin everything that does intrusion analysis, as well as how intrusion analysis is synthesized and used for network defense and mitigation [21]. However, its greatest contribution is that it ultimately applies scientific rigor and principles of measurement, testability, and respectability to the domain, allowing intrusion analysis to be more effective, efficient, and accurate, leading to faster, more effective, and efficient mitigation to defeat adversaries.

3.3 MITRE ATT&CK

MITRE is a nonprofit organization that works in the public interest in federal, state, and local governments, as well as in industry and academia. Likewise, it contributes to different areas such as artificial intelligence, intuitive data science, quantum information science, health informatics, space security, political and economic experience, reliable autonomy, threat exchange, and cyber resilience [22]. Likewise, this organization developed a methodology called ATT&CK, which is used as a basis for the development of specific threat models in the private sector, in the government, and in the community of cybersecurity products and services that contain known behaviors of attackers or adversaries, better known as advanced persistent threats (APT), which are organized attack groups with a large amount of resources and advanced techniques that allow them to carry out complex attacks [23], and in turn, research related to this methodology in critical infrastructures has been presented [24]. For example, they often stay on an organization's network and obtain information before proceeding to the next phases of their attack. ATT&CK focuses on how external adversaries engage and operate within computer information networks (Table 4).

This model originated from a project that aimed to document and categorize postengagement adversary tactics, techniques, and procedures (TTP) against Microsoft Windows systems to improve the detection of malicious behavior. [25].

Table 4 MITRE ATT&CK core components [25]

MITRE ATT&CK core components	
Tactics	Describe the tactical objectives during an attack
Techniques	Describe the means by which adversaries achieve tactical objectives
Sub-techniques	Describe more specific means by which adversaries achieve tactical objectives at a lower level than techniques
Procedures	They are the specific implementations that the adversaries have used for techniques or sub-techniques

Fig. 4 MITRE ATT&CK Enterprise Matrix [26]

Since then, it has grown to include other operating systems, as well as other areas such as mobile devices, cloud-based systems, and industrial control systems. The basis of ATT&CK is the set of techniques and sub-techniques that represent actions that opponents can perform to achieve objectives. These objectives are represented by the categories of tactics to which the techniques and sub-techniques belong. The relationship between tactics, techniques, and sub-techniques can be visualized in the ATT&CK Matrix (Fig. 4).

MITRE ATT&CK offers cybersecurity analysts a common language to structure, compare, and analyze CTI for any organization that wants to move toward an informed defense on existing or emerging threats, as it includes information on malware, tools, TTP, business techniques, behavior, and other indicators associated with threats.

3.4 MITRE Shield

Based on the ATT&CK methodology, a new methodology called MITRE Shield was developed, which is based on the implementation of the concept of active cybernetic defense, which aims to carry out cyberdefensive actions until being able to deceive the adversary, having an active participation with him to study and learn more about the tactics and techniques used to generate a CTI and prepare for future threats. The Shield matrix consists of the following components (Table 5).

Within Shield, there is also a matrix of tactics that denotes what the defender is trying to achieve through columns and techniques, which describe how the defense achieves the tactics. However, those terms have been made to fit the domain of Active Cyber Defense. These tactics within MITRE Shield must be taken into account as a strategy of each planned active defense operation in order to respond to any intrusion from an adversary or threat. Given this, it is necessary to develop the techniques described within Shield to implement security controls in an operational environment. MITRE found that a single technique can be compatible with several different tactics, and for any tactic, there are multiple techniques that can be used. In addition, it has a section where ATT&CK tactics and techniques are mapped so that defenders can have the applicable active defense information, including the presented opportunity space, the active defense technique that will be implemented, and the use case for that implementation (Fig. 5).

The combination of ATT&CK and Shield methodologies can help defenders deepen their understanding of the opponent's behavior and engagements and suggest ways the defender can implement an active defense strategy. [28]. Therefore, the goal of Shield is that defenders can take advantage of the tactics and techniques of this methodology to better create, implement and operate their active defense solutions, showing how the defensive side of Shield to align with ATT&CK, with the main objective that organizations and their defenders can take advantage of both strategies to maximize their defensive efforts and be able to generate a more solid strategy that allows them to learn from adversaries while defending against them.

Table 5 MITRE Shield core components [27]

MITRE Shield core components	
Tactics	They are abstract goals of the defenders.
Techniques	They are general actions that a defender can take that can have several different tactical effects depending on how they are implemented.
Procedures	They are implementations of a technique.
Opportunity	They describe high-level active defense possibilities that are introduced when attackers employ their techniques.
Use Cases	High-level descriptions of how a defender might do something to take advantage of the opportunity presented by the attacker's action

Channel	Collect	Contain	Detect	Disrupt	Facilitate	Legitimize	Test
Admin Access	API Monitoring	Admin Access	API Monitoring	Admin Access	Admin Access	Application Diversity	Admin Access
API Monitoring	Application Diversity	Baseline	Application Diversity	API Monitoring	Application Diversity	Burn-In	API Monitoring
Application Diversity	Backup and Recovery	Decoy Account	Behavioral Analytics	Application Diversity	Behavioral Analytics	Decoy Account	Application Diversity
Decoy Account	Decoy Account	Decoy Network	Decoy Account	Backup and Recovery	Burn-In	Decoy Content	Backup and Recovery
Decoy Content	Decoy Content	Detonate Malware	Decoy Content	Baseline	Decoy Account	Decoy Credentials	Decoy Account
Decoy Credentials	Decoy Credentials	Hardware Manipulation	Decoy Credentials	Behavioral Analytics	Decoy Content	Decoy Diversity	Decoy Content
Decoy Network	Decoy Network	Isolation	Decoy Network	Decoy Content	Decoy Credentials	Decoy Network	Decoy Credentials
Decoy Persona	Decoy System	Migrate Attack Vector	Decoy System	Decoy Credentials	Decoy Diversity	Decoy Persona	Decoy Diversity
Decoy Process	Detonate Malware	Migrate Compromised System	Detonate Malware	Decoy Network	Decoy Network	Decoy Process	Decoy Network
Decoy System	Email Manipulation	Network Manipulation	Email Manipulation	Detonate Malware	Decoy Persona	Decoy System	Decoy Persona
Detonate Malware	Network Diversity	Security Controls	Hunting	Email Manipulation	Decoy System	Network Diversity	Decoy System
Migrate Attack Vector	Network Monitoring	Software Manipulation	Isolation	Hardware Manipulation	Network Diversity	Pocket Litter	Detonate Malware
Migrate Compromised System	PCAP Collection		Network Manipulation	Isolation	Network Manipulation		Migrate Attack Vector
Network Diversity	Peripheral Management		Network Monitoring	Migrate Compromised System	Peripheral Management		Network Diversity
Network Manipulation	Pocket Litter		PCAP Collection	Network Manipulation	Pocket Litter		Network Manipulation
Peripheral Management	Protocol Decoder		Pocket Litter	Security Controls	Security Controls		Peripheral Management
Pocket Litter	Security Controls		Protocol Decoder	Standard Operating Procedure	Software Manipulation		Pocket Litter
Security Controls	System Activity Monitoring		Standard Operating Procedure	User Training			Security Controls
Software Manipulation	Software Manipulation		System Activity Monitoring	Software Manipulation			Software Manipulation
			User Training				
			Software Manipulation				

Fig. 5 The Shield Matrix [28]

4 Threat Intelligence Platform (TIP)

The techniques and results that CTI has provided in recent years have gained a great deal of attention in cybersecurity communities as a way to forecast potential threats and reduce attack detection time in terms of supply chain processes. death, as well as the use of open-source intelligence information (OSINT) is becoming a fundamental approach to gain awareness about cybersecurity threats; however, to process the large amount of information is usually one of the most important challenges in this area for defenders. Therefore, cybersecurity researchers and analysts have opted for the use of intelligence tools and platforms to be able to manage all threat analysis tasks in an orderly manner. A threat intelligence platform (TIP) helps organizations aggregate, correlate, and analyze threat data from multiple sources in real time and is comprised of several core features that enable organizations to implement an intelligence-driven approach to security to support defensive actions. The main purpose is to help organizations understand risks and protect against a variety of threat types that are likely to affect their environments collected from different intelligence sources (Table 6).

A TIP automatically analyzes the content of the threat indicators and the relationships between them to enable the production of useful, relevant, and timely threat intelligence from the data collector. This analysis allows the identification of tactics, techniques, and procedures of the threat actors or TTPs. These platforms can be a cloud or on-premises system to facilitate threat data management from a variety

Table 6 Threat Intelligence Platforms stages

Threat intelligence platforms stages	
Collect	A threat intelligence platform collects and aggregates multiple data formats for multiple sources, including formats such as STIX, CSV, XML, email, and various intelligence sources.
Correlate	The threat intelligence platform enables organizations to automatically begin to analyze correlation and pivot on data so that actionable intelligence on who, why, and how again of an attack can be obtained on the blocking measures introduced.
Enrichment and contextualization	A threat intelligence platform must be able to automatically enhance or allow threat intelligence analysts to use third-party threat analysis applications to enrich the data collected in an investigation.
Analyze	Automatically analyzes the content of threat indicators and the relationships between them to enable the production of useful, relevant, and timely threat intelligence from the data collector.
Integrate	Platform data must find a way back to the security tools and products an organization uses to enable process automation and communication.
Act	Integrated processes and workflows accelerate collaboration within the broader communications and security team, such as intelligence sharing and analytics organizations.

of existing security tools, such as SIEM, Firewall, API, Terminal Management Software, or Intrusion Prevention System (IPS). Investigations and papers have been presented where some threat intelligence exchange platforms are evaluated [29] [30], where users from the IT community and other communities in general can share their incident information in a trusted environment, such as Malware Information Sharing Platform (MISP) [31], which allows us to share, store, and correlate indicators of compromise of targeted attacks, threat intelligence, financial fraud information, and vulnerability information [32]. The malware information exchange platform can be accessed from different interfaces, such as a web interface (for analysts or incident handlers) or via a ReST API (for systems that push and pull IOCs). The inherent goal of MISP is to be a robust platform that ensures smooth operation when revealing, maturing, and exploiting threat information.

4.1 T-Pot: Platform Based on Deception Techniques

Currently, there are tools and/or platforms whose main objective is to be able to simulate a productive environment or service to maintain active communication with the adversary, commonly known as honeypots, which allows defenders to collect more information about the TTPs to be able to generate CTI. There are different honeypot platforms on the market, especially by some cybersecurity manufacturers, such as TrapX Security [33], Attivo Networks [34] and Fortinet [35], that help to

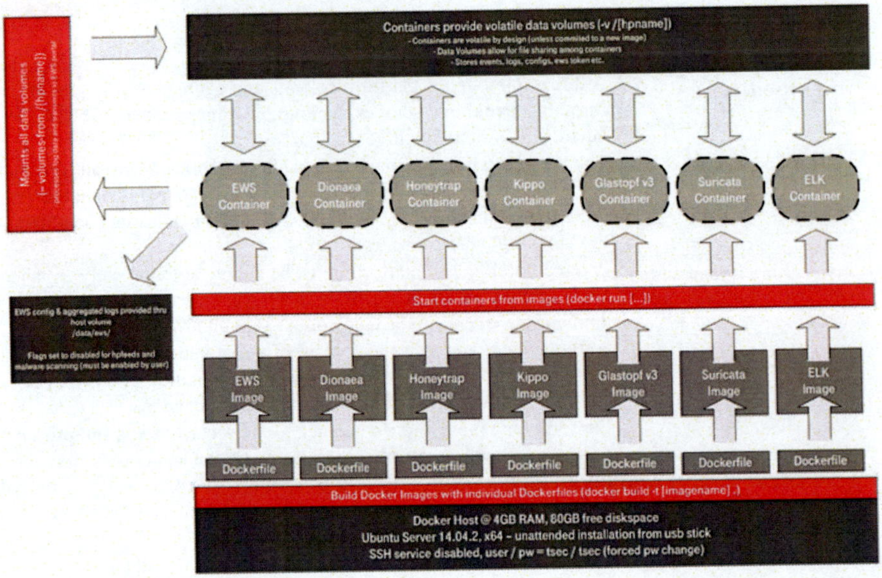

Fig. 6 T-Pot arquitecture [38]

detect external and internal threats, and studies reveal that two-thirds of the incidents found were from external actors, while the remaining third involved internal actors [36]. There are some investigations related to platforms that have multi-honeypot systems, which have focused on the investigation and relevance of the data collected through attacks. [37]. T-Pot is based on a vanilla ISO image of Ubuntu 14.04.02, which is heavily dependent on docker and docker-compose. The goal proposed by T-Pot is to create a system whose full TCP network range, as well as some important UDP services, acts as decoys for adversaries, in order to forward all incoming attack traffic to the honeypot sensors. more suitable for interacting and processing said information [38] (Fig. 6).

The project provides multiple coupled honeypots and a large number of preinstalled research tools, such as ELK, which provides a search engine and analytics, as well as an interface for data visualization. [39], Spiderfoot, which allows to perform Footprinting tasks, by acting as an aggregator of a multitude of sources, on which it allows a simple and fast search by having its own web interface [40], Cyberchef—a web application for data encryption, encoding, compression, and analysis [41]; Suricata—capable of real-time intrusion detection (IDS); online intrusion prevention (IPS); network security monitoring (NSM); and offline pcap processing [42], among others (Table 7).

Table 7 T-Pot honeypots [38]

T-Pot honeypots	
adbhoney	Low interaction honeypot designed for Android Debug Bridge over TCP/IP
ciscoasa	A low interaction honeypot for the Cisco ASA component capable of detecting CVE-2018-0101, a DoS, and remote code execution vulnerability.
citrixhoneypot	Detect and log CVE-2019-19781 scan and exploitation attempts.
conpot	Conpot is a low interactive server-side Industrial Control Systems honeypot designed to be easy to deploy, modify, and extend.
cowrie	Cowrie is a medium to high interaction SSH and Telnet honeypot designed to log brute force attacks and the shell interaction performed by the attacker.
dicompot	A Digital Imaging and Communications in Medicine (DICOM) honeypot.
dionaea	Dionaea is meant to be a nepenthes successor, embedding python as scripting language, using libemu to detect shellcodes, supporting ipv6 and tls.
elasticpot	This is a honeypot simulating a vulnerable Elasticsearch server opened to the internet.
glutton	Glutton provide SSH and a TCP proxy. SSH proxy works as a MITM between attacker and server to log everything in plain text.
heralding	Simple honeypot that collects credentials of the following protocols: ftp, telnet, ssh, rdp, http, https, pop3, pop3s, imap, imaps, smtp, vnc, postgresql, and socks5.
honeypy	A low interaction honeypot with the capability to be more of a medium interaction honeypot.
honeysap	A low-interaction research-focused honeypot specific for SAP services.
honeytrap	A network security tool written to observe attacks against TCP or UDP services.
ipphoney	This is a honeypot simulating a printer that supports the Internet Printing Protocol and is exposed to the internet.
mailoney	SMTP honeypot.
medpot	HL7/FHIR honeypot.
rdpy	Remote Desktop Protocol in twisted python.
snare	A web application honeypot sensor.
tanner	Decides how SNARE should respond to the client.

5 Experimentation and Results

In this investigation, we deployed T-Pot in the public cloud of Amazon Web Services (AWS), with the main objective of exposing the available honeypots of the T-Pot platform to the internet, in order to be able to collect real threats that are mostly automated in cyberspace, which will allow us to collect some of the adversaries' TTPs and analyze the information to model it with the methodologies described in the document through case studies, with the purpose of being able to identify how each of the models facing a real cyberattack and how they complement each other. The following is the data collected by the platform for 20 days and the characteristics and resources used to deploy the T-Pot platform on AWS (Fig. 7 and Table 8).

Fig. 7 Detected attacks on T-Pot honeypots at AWS after 20 days

Table 8 AWS-arranged resources to deploy T-Pot

T-Pot resources	
RAM	t2.xlarge (16Gbps)
vCPUs	4
OS	Linux/Debian 10
Days Active	20
Public Ip	Yes

5.1 Case Study

In this section, we analyze the data collected by the honeypots exposed to the internet to analyze the Cyber Kill Chain, ATT&CK, and Shield and Diamond Models, in order to understand some of the phases and TTP's carried out by the adversaries to achieve their objectives within the infrastructure deployed.

The case study that arises involves a victim who deploys a service with SSH, or Secure Shell, makes it very easy to access remote servers while keeping them safe from hackers. The adversaries used mass scanning techniques to detect known vulnerabilities in the exposed systems. Additionally, adversaries performed email creation using techniques such as social engineering and phishing based on previous GeoIP scans to decide the language to use. Additionally, adversaries generated an exploit to run a webshell as malware from a previously infected website. Different attempts to access the platforms by protocols such as SSH and RDP from IPs classified as malicious were detected. Some of these IPs managed to access the systems through the use of techniques such as Brute Force and dictionary attacks.

Once the adversary entered the system, command executions were detected for downloading files through a shellcode to a URL classified as malicious by malware content. After downloading, command executions were detected for downloading malware files from an IP classified as malicious. Once the malicious domains/IPs and the malware downloaded by the adversary were detected, we proceeded to review open intelligence sources such as Virus Total to search for the IPs or domains related to the installed malware. The results of the investigations carried out with the methodologies described in the document are presented below. The Cyber Kill Chain Model helps us understand each of the phases of the attack chain carried out by the adversaries and guides us to be able to start searching for the relevant events in each of the phases (Table 9).

With the information collected by the honeypot platform and the Cyber Kill Chain Model, it is possible to delve much more into the tactics and techniques used by the adversaries to understand their objectives and operations through the use of MITRE ATT&CK (Table 10).

Once the TTPs of the adversaries are modeled in ATT&CK, we can use MITRE Shield to make a map of the TTPs found in ATT&CK used by the adversaries to carry out an active defense strategy. For example, initial access to the exposed systems was done through a remote exposed system through a valid account. Given this, Shield gives us the possibility of generating an active defense strategy with the opportunity to validate if the adversary already has credentials from one or more accounts valid for any network system by using a decoy or honeypot to collect more information on the TTPs used by the adversary as we have been doing during this investigation (Table 11).

With Diamond Model, we use a victim-centered approach (honeypot) in order to reveal the connection between the adversary's related elements, such as infrastructure and capabilities organized in events (Table 12 and Fig. 8).

6 Conclusions

The conventional incident response process is generally initiated after a successful exploitation phase is executed, causing defenders to be inherently disadvantaged and their response too late. In this research, we have demonstrated the effective use of the most used methodologies in the CTI field for the detection, mitigation, and generation of CTI during the analysis and investigation process of real TTPs collected using a platform with deception techniques; it was found that these methodologies complement each other. In which, the Cyber Kill Chain Model helps us understand the phases of an adversary's attack; therefore, we obtain an initial guide for defenders on where to focus their resources, and in turn, it helps to feed the MITRE ATT&CK model, which gives us the opportunity to list and understand what TTPs are used by adversaries in each of their attack phases and provide the necessary information for the generation of an active defense strategy with MITRE Shield. Therefore, Shield is responsible for guiding us on how to take advantage of

Table 9 Case study results with Cyber Kill Chain Model

Phase	Procedure	Indicators
Reconnaissance	ip_rep.keyword: mass scanner	*Suricata Signature:* ET SCAN NMAP -sS window 1024 ET POLICY RDP connection request ET DOS Microsoft RDP Syn then Reset 30 Second DoS Attempt ET SCAN Potential SSH Scan
Weaponization	Webshell creation Previous Infected URL data: password data: covid	SSH Dictionary Attack detected Phishing emails detected
Delivery	message: login message: success eventid: cowrie.login.success src_port: 22	*Attacker IP:* 206[.]189[.]50[.]126, 46[.]101[.]156[.]22, 185[.]153[.]199[.]182 *Message Log:* login attempt / succeeded
Exploitation	message: download input: wget	*Downstream URL:* http://104[.]168[.]195[.]213/Thorbins.sh http://212[.]73[.]150[.]134/NoHomobins.sh *Exploited vulnerabilities:* CVE-2001-0540, CVE-2012-0152, CVE-2019-0708
Installation	eventid: cowrie.command.input input: wget input: run input: 104[.]168[.]195[.]213	*Malware/Trojan:* 48251b80567080237382l564eb9c7056703a6822d4e025c790d3acce0776c7fa *Malware/Downloader:* a2ef7e6b666d570dd6e26cddf4d4fd7f *Executed commands:* cd /tmp || cd /run || cd /; wget http:// 104[.]168[.]195[.]213/Thorbins.sh; chmod 777 Thorbins.sh; sh Thorbins.sh; tftp 104[.]168[.]195[.]213-c get Thortftp1.sh; chmod 777 Thortftp1.sh; sh Thortftp1.sh; tftp -r Thortftp2.sh -g 104[.]168[.]195[.]213; chmod 777 Thortftp2.sh; sh Thortftp2.sh; rm -rf Thorbins.sh Thortftp1.sh Thortftp2.sh; rm -rf *
C2	Se realizaron busquedas en Virus Total y Spiderfoot con los hashes del malware y las IP's encontradas	23[.]47[.]207[.]24:80 (TCP) 184[.]28[.]221[.]115:80 (TCP) 23[.]47[.]206[.]49:443 (TCP) 17[.]249[.]25[.]246:443 (TCP) 17[.]142[.]169[.]200:443 (TCP) 17[.]253[.]21[.]208:443 (TCP)
Actions on objectives	N/A	N/A

Table 10 Case study results with MITRE ATT&CK

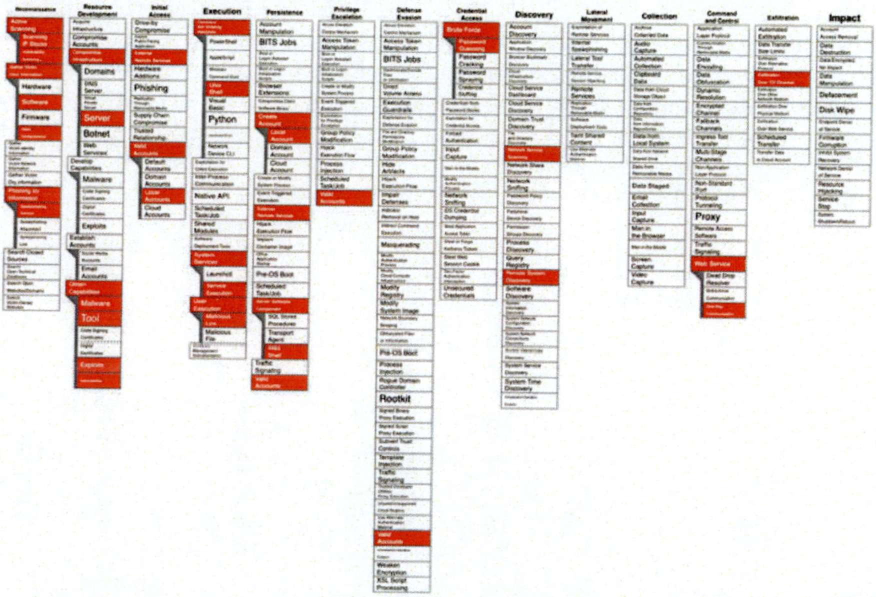

Table 11 Case study results with MITRE Shield

Channel	Collect	Contain	Detect	Disrupt	Facilitate	Legitimize	Test
Admin Access	API Monitoring	Admin Access	API Monitoring	Admin Access	Admin Access	Application Diversity	Admin Access
API Monitoring	Application Diversity	Baseline	Application Diversity	Application Diversity	Application Diversity	Burn-In	API Monitoring
Application Diversity	Backup and Recovery	Decoy Account	Behavioral Analytics	Backup and Recovery	Behavioral Analytics	Decoy Account	Application Diversity
Decoy Account	Decoy Account	Decoy Network	Decoy Account	Baseline	Burn-In	Decoy Content	Backup and Recovery
Decoy Content	Decoy Content	Detonate Malware	Decoy Content	Behavioral Analytics	Decoy Account	Decoy Credentials	Decoy Account
Decoy Credentials	Decoy Credentials	Hardware Manipulation	Decoy Credentials	Decoy Content	Decoy Content	Decoy Diversity	Decoy Content
Decoy Network	Decoy Network	Isolation	Decoy Network	Decoy Credentials	Decoy Credentials	Decoy Network	Decoy Credentials
Decoy Persona	Decoy System	Migrate Attack Vector	Decoy System	Decoy Network	Decoy Diversity	Decoy Persona	Decoy Diversity
Decoy Process	Detonate Malware	Network Manipulation	Email Manipulation	Email Manipulation	Decoy Persona	Decoy Process	Decoy Network
Decoy System	Email Manipulation	Security Controls	Hunting	Hardware Manipulation	Decoy System	Decoy System	Decoy Persona
Detonate Malware	Network Diversity	Software Manipulation	Isolation	Isolation	Network Diversity	Network Diversity	Decoy System
Migrate Attack Vector	Network Monitoring		Network Manipulation	Network Manipulation	Network Manipulation	Pocket Litter	Detonate Malware
Network Diversity	PCAP Collection		Network Monitoring	Security Controls	Peripheral Management		Migrate Attack Vector
Network Manipulation	Peripheral Management		PCAP Collection	Standard Operating Procedure	Pocket Litter		Network Diversity
Peripheral Management	Protocol Decoder		Pocket Litter	User Training	Security Controls		Network Manipulation
Pocket Litter	Security Controls		Protocol Decoder	Software Manipulation	Software Manipulation		Peripheral Management
Security Controls	System Activity Monitoring		Standard Operating Procedure				Pocket Litter
Software Manipulation	Software Manipulation		System Activity Monitoring				Security Controls
			User Training				Software Manipulation
			Software Manipulation				

the adversaries' TTPs to actively defend ourselves and collect more information in order to actively generate CTI to complement ATT&CK and the Diamond Model. Finally, the Diamond Model powered by Kill Chain, ATT&CK, and Shield, is very useful to list the capabilities and infrastructure of the adversaries, with the main objective of understanding and documenting their attack methodology to generate CTI, which can be alienated by Shield constantly.

Understanding how a cyberattack can benefit a security team in our organization can benefit the cybersecurity community by encouraging defenders to collect data on adversaries to increase the knowledge base of TTPs, facilitating the selection

Table 12 Case study results with Diamond Model

Event	Description	Phase	Methodology	Infrastructure	Capability
1	The adversaries used mass scanning techniques to detect known vulnerabilities of the exposed systems.	Reconnaissance	Active scanning	Bad IPs reputations	Massive scanners tools
2	Creation of Phishing Emails based on the previous GeoIP scans to decide the language to use.	Resource development	Spear phishing service	N/A	Email
3	The adversaries could create or obtain a malicious payload or exploit to execute a webshell as a downloader/trojan malware.	Resource development	Compromise Infrastructure Obtain Capabilities	N/A	Malicious payload/malware
4	Different attempts to access the platforms by protocols such as RDP from IPs classified as malicious were detected.	Initial access	External Remote Access	206[.]189[.]50[.]126 46[.]101[.]156[.]22	Automated scripts/tools
5	Different attempts to access the platforms by protocols such as SSH from IPs cataloged as malicious were detected.	Initial access	External Remote Access Valid Accounts	206[.]189[.]50[.]126 46[.]101[.]156[.]22 185[.]153[.]199[.]182	Automated scripts/tools and dictionaries
6	Some of these IPs managed to access the systems through the use of techniques such as Brute Force and dictionary attacks.	Credential access	External Remote Access Valid Accounts	46[.]101[.]156[.]22	Brute force attack
7	Once the adversary entered the system, command executions were detected for downloading files through a shellcode to a URL classified as malicious by malware content.	Execution	Malicious link	http://104[.]168[.]195[.]213/Thorbins.sh http://212[.]73[.]150[.]134/NoHomobins.sh	Service execution Vulnerable exposed server
8	Command to add, modify, and give execution permissions and deletion of events were detected for the downloading of malicious files.	Persistence	Web shell	http://104[.]168[.]195[.]213/Thorbins.sh	Web shell Malicious payload/malware
9	Once the malicious domains/IPs were detected and the malware downloaded by the adversary, open intelligence sources such as Virus Total were reviewed to search for the IPs or domains related to the installed malware.	Command and control	Web service	23[.]47[.]207[.]24:80 184[.]28[.]221[.]115:80 23[.]47[.]206[.]49:443 17[.]249[.]25[.]246:443 17[.]142[.]169[.]200:443 17[.]253[.]21[.]208:443	Vulnerable exposed server

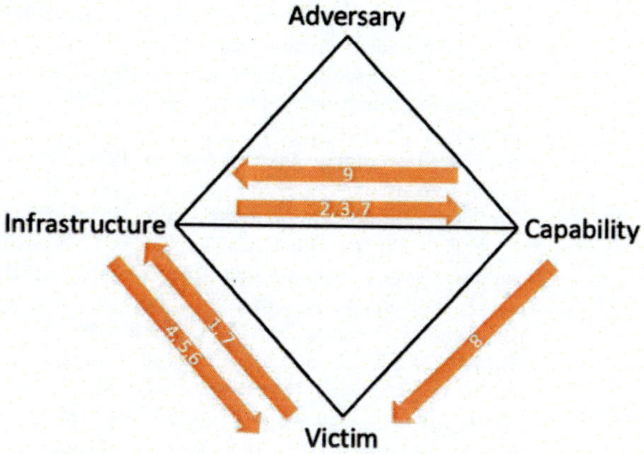

Fig. 8 Case study results with Diamond Model

of defense measures. If defenders implement countermeasures faster than their opponents evolve, they maintain a tactical advantage.

References

1. N. Gilbert, «Number of Email Users Worldwide 2020: Demographics & Predictions,» Finances Online, [En línea]. Available: https://financesonline.com/number-of-email-users/. [Último acceso: 27 10 2020]
2. I.L. Stats, «Internet Live Stats,» [En línea]. Available: https://www.internetlivestats.com/one-second/#email-band. [Último acceso: 27 10 2020]
3. Interpol, «Business Email Compromise Fraud,» [En línea]. Available: https://www.interpol.int/en/Crimes/Financial-crime/Business-Email-Compromise-Fraud. [Último acceso: 27 10 2020]
4. R.E.T. Landscape, «ENISA Threat Landscape 2020 – Ransomware,» April 2020. [En línea]. Available: https://www.enisa.europa.eu/publications/ransomware. [Último acceso: 27 10 2020]
5. Gartner, «How Gartner Defines Threat Intelligence,» Gartner, 23 02 2016. [En línea]. Available: https://www.gartner.com/en/documents/3222217/how-gartner-defines-threat-intelligence. [Último acceso: 27 10 2020]
6. R. Leszczyna, M.R. Wróbel, Threat intelligence platform for the energy sector. Softw. Pract. Exp. **49**(8), 1225–1254 (2019)
7. Y.A.R.V. Creado, Active cyber defence strategies and techniques for banks and financial institutions. J. Financ. Crime **27**(3), 771 (2020)
8. E. A. B. T. a. J. H. N. Moustafa, «A New Threat Intelligence Scheme for Safeguarding Industry 4.0 Systems,» *IEEE Access,* vol 6, 2018
9. D.C. Ahlberg, *The Threat Intelligence Handbook* (CyberEdge Group, 2019)
10. M. &. R. S. &. A. (. D. A. &. R. Y. Abu, «Cyber threat intelligence – Issue and challenges,» Indonesian J. Electr. Eng. Comput. Sci. 2018
11. Fortinet, «Threat Intelligence at Machine Speed,» [En línea]. Available: https://www.fortinet.com/fortiguard/labs. [Último acceso: 06 11 2020]
12. K. O. &. C. Doerr, «Cyber Threat Intelligence: A Product Without a Process?,» Int. J. Intell.Count. Intell., 2020

13. Intelligence-Driven Computer Network Defense Informed by Analysis of Adversary Campaigns and Intrusion Kill Chains, Lockheed Martin Corporation, «LM-White-Paper-Intel-Driven-Defense,» [En línea]. Available: https://www.lockheedmartin.com/content/dam/lockheed-martin/rms/documents/cyber/LM-White-Paper-Intel-Driven-Defense.pdf. [Último acceso: 28 08 2020]
14. SANS, «Leveraging the Human to Break the Cyber Kill Chain,» SANS, 2016. [En línea]. Available: https://www.sans.org/security-awareness-training/blog/leveraging-human-break-cyber-kill-chain. [Último acceso: 02 09 2020]
15. T. &. R. A. Yadav, «Technical Aspects of Cyber Kill Chain,» *Third International Symposium on Security in Computing and Communications,* 2015
16. T.D.A.B. Dargahi, A Cyber-Kill-Chain based taxonomy of crypto-ransomware features. J. Comput. Virol. Hack. Tech., 277–309 (2019)
17. B.I.A.A.M.S.T.S.B. Junaidu, Proposed framework for effective detection and prediction of advanced persistent threats based on the cyber kill chain. Sci. Pract. Cyber Secur. J. **3**(3) (2019)
18. lockheedmartin, «Applying Cyber Kill Chain® Methodology to Network Defense,» lockheedmartin, 2015. [En línea]. Available: https://www.lockheedmartin.com/content/dam/lockheed-martin/rms/documents/cyber/Gaining_the_Advantage_Cyber_Kill_Chain.pdf. [Último acceso: 02 09 2020]
19. A.P.C.B. Sergio Caltagirone, «The Diamond Model of Intrusion Analysis,» 2013. [En línea]. Available: https://apps.dtic.mil/dtic/tr/fulltext/u2/a586960.pdf. [Último acceso: 03 09 2020]
20. Q. M. A. N. I. A. A. C. a. J. D. Hamad AL-Mohannadi, «Cyber-Attack Modeling Analysis Techniques: An Overview,» de *4th International Conference on Future Internet of Things and Cloud Workshops*, United Kingdom, Warwickshire, 2016
21. Q. M. A. N. I. A. A. C. a. J. D. H. Al-Mohannadi, «Cyber-Attack Modeling Analysis Techniques: An Overview,» *IEEE 4th International Conference on Future Internet of Things and Cloud Workshops,* 2016
22. MITRE, «Corporate Overview,» [En línea]. Available: https://www.mitre.org/about/corporate-overview. [Último acceso: 04 11 2020]
23. MITRE, «Groups,» MITRE, [En línea]. Available: https://attack.mitre.org/groups/. [Último acceso: 04 11 2020]
24. T. A. J. C. P. M. a. S. N. G. G. R. Kwon, «Cyber Threat Dictionary Using MITRE ATT&CK Matrix and NIST Cybersecurity Framework Mapping,» de *Resilience Week (RWS)*, Salt Lake City, 2020
25. MITRE ATT&CKÒ: Design and Philosophy, «MITRE,» [En línea]. Available: https://attack.mitre.org/docs/ATTACK_Design_and_Philosophy_March_2020.pdf. [Último acceso: 04 11 2020]
26. MITRE, «Enterprise Matrix,» [En línea]. Available: https://attack.mitre.org/matrices/enterprise/. [Último acceso: 04 11 2020]
27. MITRE, «About Shield's structure and terminology,» [En línea]. Available: https://shield.mitre.org/resources/getting-started. [Último acceso: 04 11 2020]
28. MITRE, «Active Defense Matrix,» [En línea]. Available: https://shield.mitre.org/matrix/. [Último acceso: 04 11 2020]
29. A. de Melo e Silva, J. J. Costa Gondim, R. de Oliveira Albuquerque y L. J. GarcÃa Villalba, «A Methodology to Evaluate Standards and Platforms within Cyber Threat Intelligence,» *Future Internet,* 2020
30. C. S.,. A. M.,. a. R. B. Clemens Sauerwein, «Threat Intelligence Sharing Platforms: An Exploratory Study of Software Vendors and Research Perspectives,» de *Internationalen Tagung Wirtschaftsinformatik*, St. Gallen, Switzerland, 2017
31. 10.1145/2994539.2994542, «MISP – The Design and Implementation of a Collaborative Threat Intelligence Sharing Platform,» *Workshop on Information Sharing and Collaborative Security,* 2016.
32. MISP, «MISP – Open Source Threat Intelligence Platform & Open Standards For Threat Information Sharing,» [En línea]. Available: https://www.misp-project.org/features.html. [Último acceso: 08 11 2020]

33. TrapX, «TrapX,» [En línea]. Available: https://trapx.com. [Último acceso: 07 11 2020]
34. Attivo Networks, «ThreatDefend® Detection & Response Platform,» [En línea]. Available: https://attivonetworks.com/product/deception-technology/. [Último acceso: 07 11 2020]
35. Fortinet, «FortiDeceptor,» [En línea]. Available: https://www.fortinet.com/products/fortideceptor. [Último acceso: 07 11 2020]
36. 2. D. B. I. Report, «Enterprise Verizon,» [En línea]. Available: https://enterprise.verizon.com/resources/reports/2019-data-breach-investigations-report.pdf. [Último acceso: 07 11 2020]
37. B.J.R.E. Sanjeev Kumar, «Multi Platform Honeypot for Generation of Cyber Threat Intelligence,» de *9th International Conference on Advanced Computing (IACC)*, 2017
38. Telekom Security, «Introduction into T-Pot: A Multi-Honeypot Platform,» 2015. [En línea]. Available: http://github.security.telekom.com/2015/03/honeypot-tpot-concept.html. [Último acceso: 07 11 2020]
39. elastic, «¿Qué es el ELK Stack?,» [En línea]. Available: https://www.elastic.co/es/what-is/elk-stack. [Último acceso: 07 11 2020]
40. spiderfoot, «Spiderfoot,» [En línea]. Available: https://www.spiderfoot.net. [Último acceso: 07 11 2020]
41. Crown Copyright 2016, «Cyberchef,» [En línea]. Available: https://gchq.github.io/CyberChef/ . [Último acceso: 07 11 2020]
42. Suricata, «Suricata,» [En línea]. Available: https://suricata-ids.org. [Último acceso: 07 11 2020]

Probabilistic Update Policy to Deal with Bandits for Staged Problems Applied to Cases

Carmen Constanza Uribe Sandoval and Luis Oliverio Chaparro Lemus

Abstract Some decision problems are represented in sequential stages within which an action is executed without knowing its effect until the action of the last stage is completed. A dynamic case management modeling and notation problem has been remodeled with these features to improve its automation. This chapter describes a bandit-based application with a probabilistic learning policy tested with simulated data and a stage graph and proposes its application in case automation. The results of the simulations and an initial model of the application as a graph of stages are presented.

Keywords Machine learning · Decision-making · Bandit models · Case · Dynamic processes

1 Introduction

Dynamic processes automation in organizations is a problem in the Business Process Management Suits—BPMS industry; for this purpose, Case Management Modeling and Notation—CMMN tools have emerged that allow the modeling and automation of cases. Business Process Management Notation—BPMN and CMMN—environments pursue the modeling and automation of all types of human activity (business) that can be executed as a process, whether determined or not. However, automation with CMMN can be improved (decrease its uncertainty) by

C. C. Uribe Sandoval (✉)
Universidad de Boyacá, Tunja, Colombia - Universidad Autónoma de Nuevo León, Monterrey, México
e-mail: ccuribe@uniboyaca.edu.co

L. O. Chaparro Lemus
Universidad de Boyacá, Tunja, Colombia
e-mail: lochaparro@uniboyaca.eu.co; http://www.uniboyaca.edu.co

F. Torres-Guerrero et al. (eds.), *2nd EAI International Conference on Smart Technology*, EAI/Springer Innovations in Communication and Computing,
https://doi.org/10.1007/978-3-031-07670-1_3

learning certain patterns after repeated execution of the case model implementation [8].

Decision-making has reached a large volume of work and research due to its application to private and corporate problems. Some researchers and scholars have made use of graphs to model decision-making situations [1] and have proposed algorithms to solve them, several of which have been studied in different undergraduate programs such as engineering [3]. On the other hand, Artificial Intelligence has been the basis for the development of programs that allow approximating recommendations that previously would only be made by a human expert [9], which, clearly, may also have applications in decision-making problems.

In Artificial Intelligence, three types of machine learning are used: supervised, unsupervised, and reinforcement learning. For the last one, the system does not learn from the information provided by an external subject; instead, it has to discover what it must learn according to the consequences of its actions. A simple technique within reinforcement learning is the one that allows solving the problem of the multi-arm bandit (MAB), where an agent must decide the action that generates the best reward from among a set of them, for which its associated value is only known until a large number of attempts have been made and the value can be deduced. This technique is approached in this chapter to solve stationary processes related to the management of probabilities associated with each action since it uses the bandit gradient model [11].

The set of actions available to the user to find the one that gives the highest reward can be modeled using a graph. Actually, in this chapter, the multi-armed bandit technique is used with some restrictions, suitable for problems that can be modeled in stages, where the individual value of the reward in each action is only known when the final objective has been achieved; that is, there is uncertainty throughout the process.

Uncertainty is one of the characteristics of dynamic business cases [5]. The issue shown above inspired this work; since a decision-maker does not have a standard procedure that ensures the sequence of actions for a case to be resolved but knows the actions eventually involved, the restrictions present between those actions and the final result when selecting a sequence of them; that is, the achievement or not of the expected objective. Some samples of decision problems appear in everyday life; they are made up of a set of activities that are executed, and where the reward is only observed after they have been executed many times, such as medical health care processes, in which only after performing a certain set of activities on a patient, can the effect of the process be established.

On the other hand, the multi-armed bandit technique has already been studied as an alternative solution to complex problems with little information, thanks to its characteristic balance between exploitation and exploration, which means good results with a lower computational cost compared to other techniques [14].

As an alternative that allows the recommendation of a tasks' sequence, each task belonging to a well-defined stage, with a low computational cost, we implemented the multi-armed bandit model of Reinforcement Learning in this research and describe it in this chapter.

2 Multi-Armed Bandit Problem

"Multi-armed bandits is a simple but very powerful framework for algorithms that make decisions over time under uncertainty" [10]. There are several types of multi-armed bandits related to the degree of complexity of the algorithms, the way they interact with the environment, and the mathematical supports that characterize them, among other aspects.

In [6], the bandit process is described as a single-armed bandit process, as a machine that has associated a sequence of states, each of them with a reward, and a state transition function that operates according to the states already visited and an independent real-valued random variable, and with a known statistical description. And, the multi-armed bandit process is described as a set of single-armed bandit processes with a controller that operates one of these machines for each time, based on a policy that it adopts to maximize the rewards received.

Thus, the accumulated benefit that is received when selecting an action "a" at different times i is a simple average of the rewards that have been accumulated when selecting action "a," since the reward will not necessarily be the same at different times, as it can be seen in formula 1 [11].

$$Q_t(a) = \frac{R_1 + R_2 + \ldots + R_{Nt(a)}}{N_t(a)}. \tag{1}$$

The same author [11] considers a gradient bandit when he learns a number preference $H_t(a)$ for each action a, but it "has no interpretation in terms of reward." This preference directly influences the probability that the action will be taken in the next time, according to the Boltzmann distribution (Eq. 2), and this is updated in each step, according to the reward and the average of all the rewards up through and including that time.

$$\pi_t(a) = \frac{e^{H_t(a)}}{\sum_{b=1}^{n} e^{H_t(b)}}. \tag{2}$$

Initially, all $H_t(a)$ preference values are equal, regardless of their value, so all actions have the same probability of being selected.

3 The Problem of Dynamic Cases

Business Process Management—BPM, according to [4], refers to the design, representation, analysis, and control of business processes and brings together a set of process management best practices with tools and information technologies. For [2], it is the integration of information technologies to improve, innovate, and manage business processes to facilitate the achievement of business objectives.

The management of processes has been a constant concern for researchers who have been searching for their optimization, specifically the Workflow, such as some of the models presented in [13]. It has gone from managing purely static processes to increasingly dynamic processes, which has given rise to new challenges for companies that generate Business Process Management Suites (BPMS) where all the technological developments that support business processes are concentrated [12], a challenge that this research aims to support.

As stated above, the term "case" arose, which in [13] is defined as a situation that may occur in the business for which the resolution procedure is not necessarily predefined. One of the emerging standards for modeling cases is Case Management Model and Notation (CMMN) that uses a set of graphic symbols, composition rules, and artifacts for this purpose; a comprehensive description of this notation is found in [7], but it is emphasized that the dotted lines around the activities make them optional, and this reveals the uncertainty that characterizes the cases.

An example of a case modeled with CMMN is shown in Fig. 1. This is the model that is discussed and analyzed later.

4 Model

In this way, decision problems are modeled where it may be necessary and sufficient to select one and only one of the nodes in each stage or level (see Fig. 2), to form a set that, in the end, will allow a planned objective to be reached or not. Each of the nodes in the graph has an associated bandit. At the end of many iterations, it is expected that the path found—the convergent path by this algorithm—will be made up of the actions that have achieved the highest preference, and therefore, it must match a route that will reach the goal with a high reward.

The probability of taking a path made up of the L nodes will be equivalent to the multiplication of the transition probabilities between its nodes, as expressed in Eq. 3 where i denotes the decision node selected at a given stage and L is the number of stages.

$$P(i_0 \rightarrow i_1, i_1 \rightarrow i_2, \ldots, i_{L-1} \rightarrow i_L) = P(i_0 \rightarrow i_1)P(i_1 \rightarrow i_2) \ldots P(i_{L-1} \rightarrow i_L). \tag{3}$$

Since the model must be adapted to problems where the reward associated with each node is not known, but rather a global reward for each complete path, the concept of preference for node selection of the Gradient bandit mentioned above is used. Here, the preference associated with node i is called v_i, and the transition probabilities between adjacent nodes are estimated from those preference values, as observed in Eq. 4, where A is the adjacency matrix of the network of bandits.

$$P(i \rightarrow j) = \frac{e^{v_j}}{\sum_k A_{i,k} e^{v_k}}. \tag{4}$$

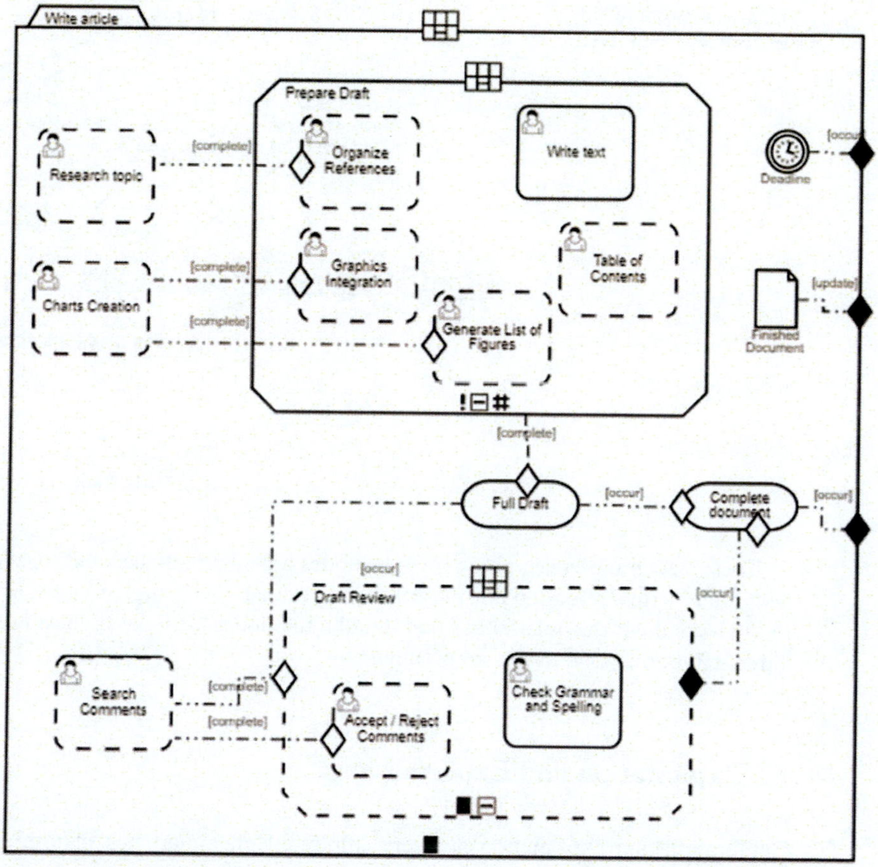

Fig. 1 CMMN diagram. Adapted from [7]

The initial values of v are zero, so the formula 4 calculates the same probability to go from one node to each of its next or neighboring nodes, that is, with a uniform probability; in this way, the decision to move from node i to another node k is totally random.

After each stage τ, the total reward is observed, and according to its sign, a positive or negative reward will be given to the preferences associated with all the nodes of the path, as Eq. 5 where $\delta \geq 0$ is a parameter that controls the learning rate.

$$v_j(\tau + 1) = v_j(\tau) \pm \delta. \tag{5}$$

Although values close to zero are expected for δ, after the implementation of the model, it became necessary to handle values with magnitudes related to the rewards of the problem being tested, as will be explained later.

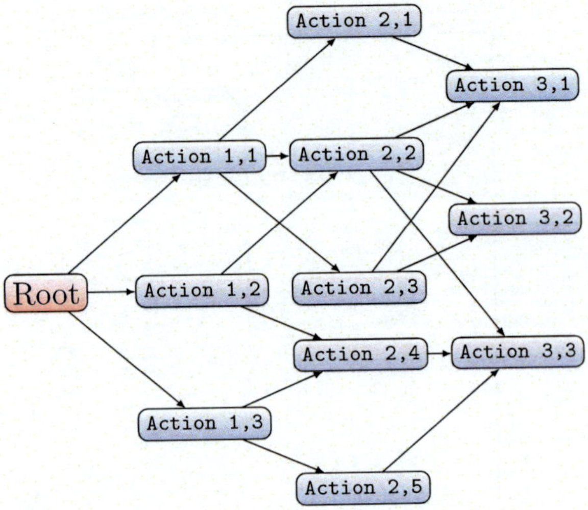

Fig. 2 A network of hidden bandits with $L = 3$ levels

These new preference values v_i of the nodes are taken into account to calculate a new route, with a selection priority for the nodes that have a higher value, according to the formula 4; this guarantees that, finally, the nodes with the highest preference value will correspond to the solution path.

5 Exploitation and Exploration

Exploration and exploitation are very important characteristics in reinforcement learning models.

Exploitation refers to the tendency of the agent to continue selecting the activity that has given the best result so far, in search of improvements. Exploration refers to the agent's ability to evaluate other activities that are not necessarily working well, looking for a better selection in the future [11].

The algorithm that is presented gives greater importance to exploration when it is beginning to learn or to know the effects of its decisions, which gradually decreases, while the importance of exploitation increases toward the end of learning. Although one of them prevails at a certain moment, there is always a percentage of probability for the other. This happens because the selection criterion is given by the transition probabilities between nodes according to Eq. 4, which with the initial values of v_i at zero guarantees uniform probabilities to pass from a node to its adjacent ones for the next stage, giving all the space to the exploration, since there is no preferred node. In each time or execution, these probabilities are updated according to the modifications made to the preference values v_i with increases or decreases δ whose magnitude has been adjusted to guarantee that a sufficient percentage of selection of alternatives that have not yet been chosen, during the first iterations.

In intermediate cases in which the values of the probabilities begin to mark their superiority over those of other transitions, there is a lower percentage of probability that other actions will be selected, increasing exploitation and decreasing exploration; but the moment will come when the nodes that make up the answer to which the search converges have almost 100% of the probability of being chosen, and it is at that moment that the answer of the recommended path is obtained.

6 Algorithmic Implementation of the Model

The generation of a graph for the tests of the learning algorithm has been implemented, whose adjacency matrix is stepped so that no node is adjacent to another of the same stage. When generating the matrix, it is also taken into account that each node must have at least one adjacent node from the next stage. The nodes of the graph correspond to the actions that are taken, and each of them is assigned a reward to calculate the gain of the path at the end. Each node is associated with a bandit that a normal distribution has a mean in its reward and standard deviation of a magnitude according to those of the rewards.

Once the gain or loss is known at the end of the iteration, the preference values of each node of that route are updated, according to the formula 5; this value is used to calculate the new transition probabilities with the formula 4, for the next iteration. It should be noted that, at the end of each stage, only the transition probabilities of the nodes involved in the selected route are affected, but that at the beginning of each iteration, all the modifications that these probabilities have undergone in the previous stages of the simulation are taken into account.

The selection for the next neighbor of each node is strongly influenced by the probabilities of transition between nodes, which give rise to the exploitation of the preferred routes, and to the exploration as explained in Sect. 5. So, at the end of the defined iterations, the simulator converges to the best route it has studied, which may be the optimal one.

Algorithm 1 shows the pseudocode of the proposed model. It knows the values of L (the number of stages), M (vector with the number of nodes of each stage), and the vector w (preference values of each node) that is initialized to zeros. Calculate n that is the total number of nodes in the graph, randomly to generate the adjacency matrix and the bandits for each node. For each iteration, the simulator calculates the transition probabilities between nodes, with which, in each stage (up to the penultimate), it selects a node and finds its neighbors and does the same in the next stage.

With the approximate bandits' values of the selected nodes, the gain at the end of the L stages is calculated. According to the value and the sign of the gain, the preference values of each node (w) are updated with which the probabilities of transition to its neighbors are calculated again for the next iteration.

Algorithm 1 L-n-bandit(L, M[L])

Calculate: n
Initialize: v[n] = 0
Generate: Ad[nxn] = Adjacent Matrix
Generate: B[n] = Real Bandits
for t = 1 **to** T **do**
    Initialize: orig = 0
    Add: orig in path
    $P(i \to j) = \frac{e^{v_j}}{\sum_k A_{i,k} e^{v_k}}$,
    for l = 1 **to** L-1 **do**
        Generate: nvz_{orig} = neighbours
        Select: $dest \in nvz_{orig}$ by $P(i \to j)$
        $orig = dest$
        Add: orig in path
    end for
    Calculate: $Gain_{path}$
    if $Gain_{path} > 0$ **then**
        v[n+1] = v[n] + δ by i $\in$ path
    else
        v[n+1] = v[n] - δ by i $\in$ path
    end if
end for

7 Numerical Experiments

The code is executed 10 consecutive times to obtain an initial estimate of its effectiveness; each time with 999 iterations, a percentage of seven out of ten executions is obtained that lead to the correct answer following the approximate generation procedure of the probability transition matrix, as can be seen in the composite Fig. 3.

Then, to estimate the performance of the algorithm and the most suitable value for the parameter δ, we take a fixed graph arranged as shown in Fig. 4, that is, with five stages and thirteen nodes, connected as shown the edges and their adjacency matrix.

This graph complies with the defined restrictions has a high density of allowed connections and in each stage has one of the values of its bandits higher than that of the other ones, to ensure that it finds the route that has the highest gain associated with it. The adjacency matrix and the bandit vector values (B) were generated as follows.

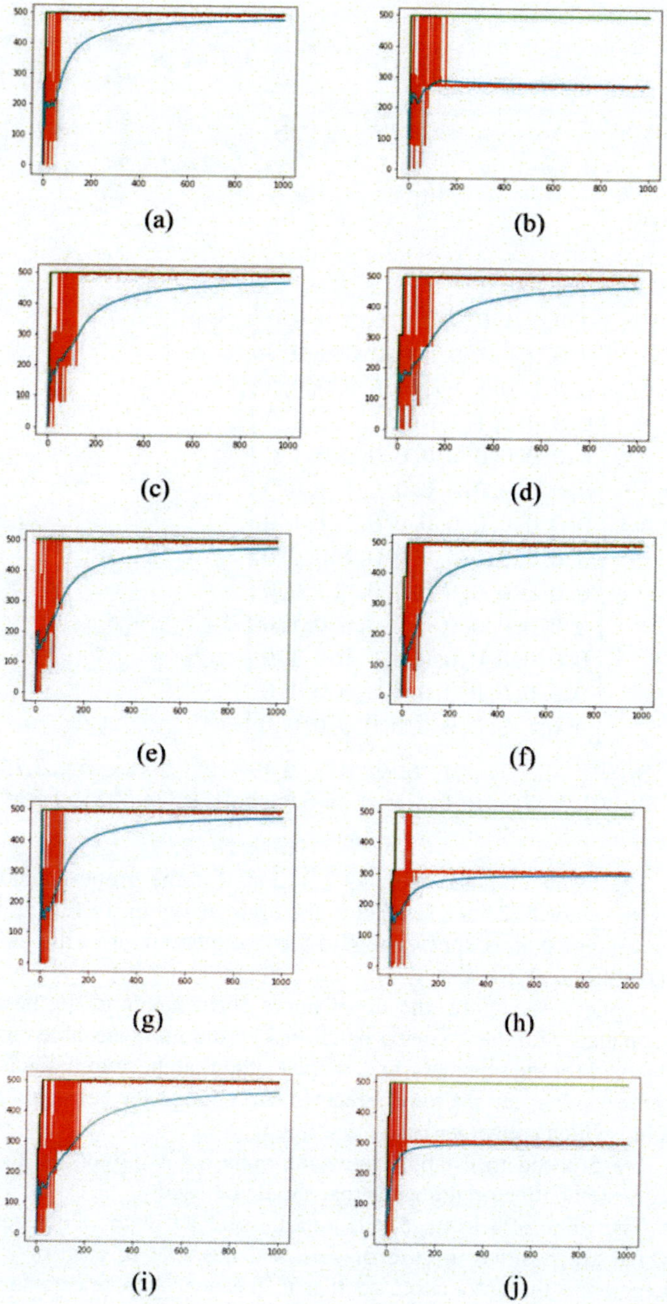

Fig. 3 Results with random graphs

Fig. 4 Model graph

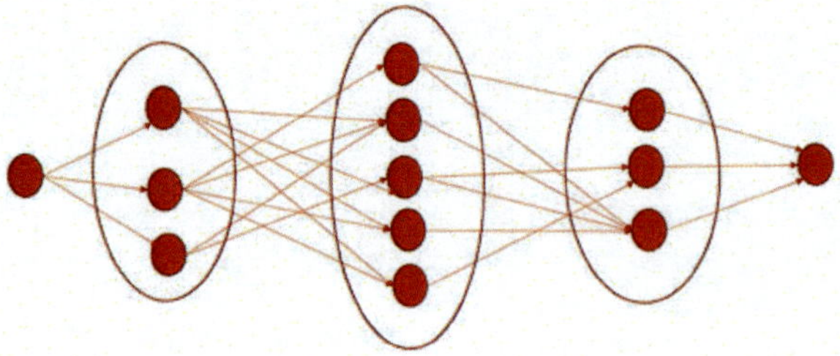

$$
A = \begin{bmatrix}
0,0,1,1,0,0,0,0,0,0,0,0,0 \\
0,0,0,0,1,0,0,0,0,0,0,0,0 \\
0,0,0,0,1,0,0,1,1,0,0,0,0 \\
0,0,0,0,1,1,0,1,1,0,0,0,0 \\
0,0,0,0,0,0,0,0,0,0,0,1,0 \\
0,0,0,0,0,0,0,0,0,1,1,0,0 \\
0,0,0,0,0,0,0,0,0,1,1,1,0 \\
0,0,0,0,0,0,0,0,0,1,0,0,0 \\
0,0,0,0,0,0,0,0,0,0,1,0,0 \\
0,0,0,0,0,0,0,0,0,0,0,1,0 \\
0,0,0,0,0,0,0,0,0,0,0,1,0 \\
0,0,0,0,0,0,0,0,0,0,0,1,0 \\
0,0,0,0,0,0,0,0,0,0,0,0,0
\end{bmatrix}
$$

$B = [-0.833e{-}01,\ 6.12e{+}02,\ 9.49e{-}02,\ 3.18e{+}00,\ 3.75e{-}01,\ 6.33e{-}01,\ 7.73e{-}01, 5.47e{-}01, 5.77e{+}02, -6.95e{-}01, 8.23e{+}02, -1.83e{+}00, -3.13e{-}01]$.

The result obtained with 99 iterations for the proposed model is the 5-stage nodes vector [0,2,4,11,12] that is the optimal vector. In Fig. 5, it can be seen how the average gain is approaching to the real gain and how the gain converges to the maximum real gain curve.

In the same figure, the upper curve corresponds to the best-tested route gain calculated with the assigned bandit value; the most unstable curve shows the gain obtained for the selected route, to the extent that the automaton is learning the value of the bandits; and the lower greenish curve shows the accumulated average of these gains, which converges to the first line.

We proceed to establish how the value of δ influences, to give an informed recommendation on the value that should be used.

The value of δ in Eq. 5 was initially selected with values between 0.1 and 0.9, which are positive in case of gain and negative in case of loss, to motivate or demotivate the automaton, so that in the next step select the nodes that receive the reinforcement. The normalized differences that were found between the real gain and each of the gains that were obtained on average in 100 iterations for each parameter's combination are shown in Table 1.

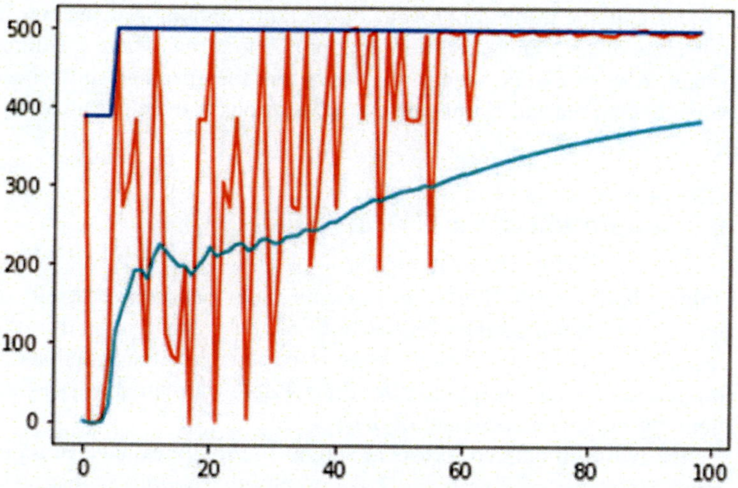

Fig. 5 Probability-weighted result

Table 1 Results with δ

	Iterations			
Deltha	100	300	600	900
1e−1	0.7146873	0.6781488	0.7089705	0.7325367
1e−2	0.6965094	0.6706974	0.6768052	0.6542384
1e−3	0.2909915	0.2335418	0.2185707	0.1505568
1e−4	0.6362933	0.3897829	0.2263654	0.1647789
1e−5	0.7059966	0.6926970	0.6712759	0.6435113

Table 2 Results with gamma

	Iterations					
Gamma	100	300	500	700	900	1000
0.00001	0.368	0.360	0.344	0.328	0.322	0.272
0.00025	0.102	0.044	0.034	0.066	0.052	0.042
0.00050	0.050	0.078	0.082	0.072	0.082	0.074
0.00075	0.080	0.092	0.086	0.090	0.078	0.090
0.00099	0.092	0.102	0.084	0.084	0.100	0.106

The difference ranges are over 15%, with some little differences for the number of iterations that are executed. Best δ values are in magnitudes of thousands as are viewed in Table 1.

To improve it, we proceed to create a self-adjusting δ, which depends on the value of the gain received by the tested route. A new variable γ is involved, which, when multiplied by the profit or reward received at the end of the path, will give the value δ for Eq. 5.

In fact, it was necessary to take values for γ of the order of ten thousandths, so that when multiplying it by the gain, which for this example reaches values close to 500 units, values of δ less than one can be obtained as viewed in Table 2.

Due to it, the results improved considerably, obtaining an average in the error for this table of 13% and of 8% eliminating the first row where the highest values are found. Also, it can be seen in this table that the errors are influenced both by the value of Gamma and by the number of iterations that are executed.

8 Adaptation of the CMMN Model

Taking into account Fig. 1, it is established that it is possible to express the information recorded there in the form of the graph proposed in this chapter. As the dotted lines that surround some activities make them optional, they must be replicated in several stages to offer the possible alternative ways to achieve the final goal: the complete document of the article.

The two well-defined sectors of actions in Fig. 1 are analyzed separately, and the graphs in stages of Figs. 6 and 7 are proposed. The idea is that once the graph of the "Prepare draft" stage is concluded, the paths are continued with the graph of the "Review draft" stage.

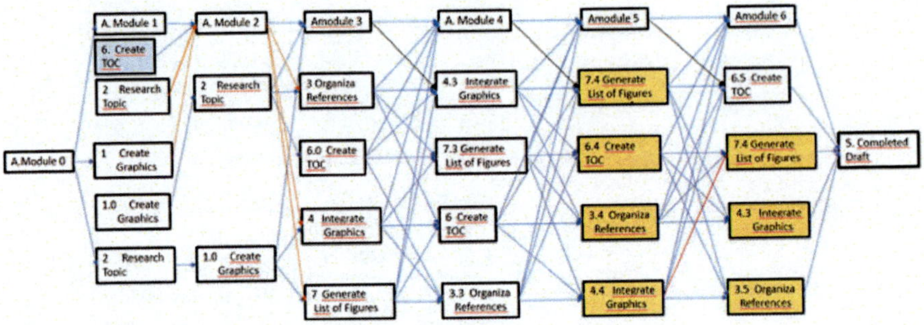

Fig. 6 Prepare draft in stages

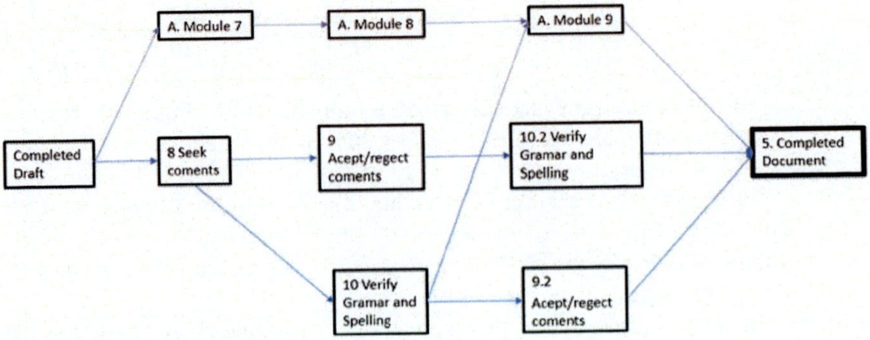

Fig. 7 Review draft in stages

In addition to the fact that it was necessary to repeat activities in each step of the graph, a null activity was also added in some stages of Fig. 6, so that in this part, a sequence with less than seven activities can be formed, even with only the mandatory activity "write text."

This is the beginning of a new experimental research work that establishes the validity of this proposal and its implementation in Case Management in some BPMS.

9 Conclusions

Various topologies of graphs are found in the literature, but the step graph proposed in this paper is novel and allows the particular modeling of some problems in which activities have to be selected in sequential times, where the result of the decisions that are taken in each stage, it will only be known at the end of an amount L of given stages.

Reinforcement learning and, in particular, the multi-armed bandit (MBA) model offer good results, with low computational cost, thanks to the importance it gives to both exploration and exploitation when searching for solutions.

The calculation of the transition probabilities between nodes, implemented in this proposal, makes it possible to recommend the best possible solution vector. The proposed model was tested with a 5-stage graph with 13 nodes and reaches its convergence in less than 100 iterations.

The BPMS industry can benefit from the proposal that was released, so it is necessary to recommend the continuity of this work in specific cases.

On the other hand, it is necessary to finish the implementation of the model in CMMN cases, making tests in several topics until generalize for any case of any topic.

The proposed algorithm for the stage problems could be improved, compared with other algorithms, and tested for more scenarios.

Bibliography

1. A. Baykasoglu, A review and analysis of "graph theoretical-matrix permanent" approach to decision making with example applications. Artif. Intell. Rev. **42**, 573–605 (2014). https://doi.org/10.1007/s10462-012-9354-y
2. T. Benedict , N. Bilodeau, P. Vtkus, M. Powell, D. Morris, M. Scarsig, D. Lee, G. Field, T. Lohr, R. Saxena, BPM CBOK Version 3.0: Guide to the business process management common body of knowledge. CreateSpace Independent Publishing Platform (2013)
3. S. Even, *Graph Algorithms* (Cambridge University Press, Cambridge, 2011)
4. K. Garimella, M. Lees, B. Williams, *Introducción a BPM para Dummies® Edición especial de Software AG* (Wiley Publishing, Inc. Indianápolis, 2008)
5. R.J. Madachy, *Software Process Dynamics* (Wiley, Hoboken, 2007)

6. A. Mahajan, D. Teneketzis, Multi-armed bandit problems, in *Foundations and Applications of Sensor Management* (Springer, Berlin, 2008), pp. 121–151
7. Object Management Group, *Case Management Model and Notation (CMMN)* (Object Management Group, Needham, 2013)
8. Object Management Group, *Case Management Model and Notation (CMMN) Version 1.1* (Object Management Group, Needham, 2016)
9. J.-C. Pomerol, Artificial intelligence and human decision making. Eur. J. Oper. Res. **99**(1), 3–25 (1997)
10. A. Slivkins, Introduction to multi-armed bandits (2019). arXiv:1904.07272
11. R.S. Sutton, A.G. Barto, *Reinforcement Learning: An Introduction* (MIT Press, Cambridge, 2018)
12. W.M.P. Van Der Aalst, Business process management demystified: a tutorial on models, systems and standards for workflow management, in *Advanced Course on Petri Nets* (Springer, Berlin, 2003), pp. 1–65
13. W.M.P. Van der Aalst, M. Weske, D. Grünbauer, Case handling: a new paradigm for business process support. Data Knowl. Eng. **53**(2), 129–162 (2005)
14. P. Zhou, J. Xu, W. Wang, Y. Hu, D.O. Wu, S. Ji, Toward optimal adaptive online shortest path routing with acceleration under jamming attack. IEEE/ACM Trans. Netw. **27**(5), 1815–1829 (2019)

Virtual Job Expo: A Practical Approach to Virtual Reality in Different Development Engines

Ordoñez Sergio, Rodríguez Iveth, and Neira-Tovar Leticia

Abstract Virtual reality (VR) is a technology that allows emulating a synthetic environment giving users the perception of being elsewhere or augmented reality that mixes the real environment with additional elements creating the illusion that the environment is a mixture of both worlds, the virtual and the real. These technologies allow people through devices to recreate sensations from visual to tactile becoming increasingly realistic, detailed, interactive, and exciting.

Virtual reality has not only served for recreational purposes but has been a fundamental part in education, in research, and in the replacement of activities that at the time entailed physical effort and threaten safety and even involve high costs or simply the provided environments it was impossible in a real environment to be present in each of them. One of the targets that this work promotes is the use of virtual reality into the workforce research. Actually, employers search for candidates who have evolved the use of new technologies to select and hire human capital. At this work, a review of two platforms that uses VR (virtual reality) in job expos is exposed, with the aim of showing its usefulness as a means of virtual rapprochement between employers and job seekers

Keywords Immersive reality · Virtual reality recruit · VR framework · Job sites · Video-games design

1 Introduction

Virtual reality has long been touted for its potential to revolutionize education, with countless advantages cited: access to remote experts, access to experiences that depend on scarce resources or limited access (e.g., going to the moon), and access

O. Sergio (✉) · R. Iveth · N.-T. Leticia
Universidad Autónoma de Nuevo León, San Nicolás de los Garza, N.L., Mexico
e-mail: sergio.ordonezg@uanl.mx; diana.rodriguezsncz@uanl.edu.mx; leticia.neiratv@uanl.edu.mx

F. Torres-Guerrero et al. (eds.), *2nd EAI International Conference on Smart Technology*, EAI/Springer Innovations in Communication and Computing,
https://doi.org/10.1007/978-3-031-07670-1_4

to experiences that are physically impossible (e.g., standing inside a module), to name a few. A new generation of consumer hardware has made this vision more affordable than ever. The interest is to understand what advantages of virtual reality in an educational context will determine when and how it will happen. The named advantages for collaborative virtual environments fall into two broad categories: in interaction with other humans and those focused on the environment. Human interaction can be novel because of who you can interact with (e.g., remote people) or how you can interact (e.g., by assuming a physical appearance). The environment can be novel because it is based on a physical location that only a few people can go to or because the experience it provides is inherently virtual (e.g., being inside a molecule) [1]. The Virtual Job Fair consists of offering the user (whether student or company) a completely virtual experience located at the exposure site, where the company can exhibit its stands to recruit the student and students can take a tour to obtain information on vacancies and get statistics.

Thanks to the appropriate measures being taken to replace school activities, this will lead to continuing innovating new information technologies and every day to reach new goals.

2 Hardware and Software Needed for Virtual Reality

For all who are willing to step into the world of virtual reality, a specific set of hardware accessories is needed to make it possible.

The set mentioned above could be comprised of:

- Head-mounted screens consisting of two small screens, one for each eye, a material used to stop light from approaching the real world, and a pair of stereo headphones with the function to give users awareness of the environment.
- Immersive rooms that represent an alternative to head-mounted screens, lists of areas that contain special projections, and tours that turn walls into exhibits. This highly advanced room also contains an array of specialized sensors that can track people inside, thus moving the projected images according to their movement.
- Data gloves that are used to give people the ability to interact with the virtual object objects that make the experience more realistic. This technology involves very sophisticated sensing strapping on ordinary gloves [2].

For the Virtual Job Fair project, it has been thought and designed to be supported by VR equipment such as Vive, Rift, Windows Mixed Reality, Daydream, Gear VR, Cardboard, Oculus Go, and 360° equipment with a rendering quality on midrange equipment up to 90 fps and high-end equipment up to 120 fps. On the software side, the use of the A-Frame platform has been implemented, which is supported by Supermedium, Firefox, Oculus Browser, Samsung Internet, Microsoft Edge, Chrome, Exokit, Safari for iOS, Chrome for Android, Firefox for iOS, and UC Browser (Fig. 1).

Fig. 1 Compatible virtual reality equipment

3 A-Frame

A-Frame is a framework for creating virtual reality (VR) experiences. A-Frame is based on HTML, which simplifies getting started. But A-Frame is not just a 3D scene graph or markup language; the core is a powerful entity-component framework that provides a declarative, extensible, and composable structure to Three.js.

Originally conceived within Mozilla and now maintained by A-Frame's co-creators within Supermedium, A-Frame was developed to be an easy yet powerful way to develop virtual reality content. As a stand-alone open-source project, A-Frame has grown into one of the largest virtual reality communities.

A-Frame is compatible with most VR headsets like Vive, Rift, Windows Mixed Reality, Daydream, Gear VR, Cardboard, and Oculus Go and can even be used for augmented reality. Although A-Frame supports the full spectrum, A-Frame aims to define fully immersive interactive VR experiences that go beyond basic 360° content, making full use of positional tracking and controllers [3] (Fig. 2).

4 Implementation of 3D

The use of 3D (three-dimensional representation of geometric data) has wide implementations both within the industry and in the research area; based on this,

Fig. 2 A-Frame framework photo credits: https://aframe.io/

the 3D department of the Faculty of Mechanical and Electrical Engineering was given the task of carrying out the design and rendering of the images of the faculty as well as the complete design of the stands that will house the different affiliated companies. This implementation is made up of several fundamental factors; among them are architecture, physics, electronics, and multimedia.

4.1 Architecture

The creation of 3D models is something that in the professional architecture area is essential to represent structures or buildings in order to visualize and have the possibility of considering whether what is planned requires any change or not.

4.2 Physics

Within this area, it is possible to create assets which can be used to emulate an object in real life, in order to carry out simulations, such as experimenting with how aerodynamic an object can be.

Fig. 3 School area rendered by the 3D department for the implementation of the Virtual Work Fair 2020

4.3 Electronics

Since 3D printing has become accessible to the specialized 3D modeling public, the generation of plastic parts has become common for electronics projects where it is necessary to generate original or unusual parts.

4.4 Multimedia

3D is a tool that is increasingly used in all fields of multimedia, certain things that may seem real many times tend to be 3D models which with techniques and special effects give the impression that they are really there, and this is highly useful for when you need to generate something that is not easy to have or generate either in real life or in 2D animation, so the use of 3D is increasingly common in advertising, cinema, and animations [4] (Figs. 3 and 4).

5 Virtual Experience

A fundamental part to implement the virtual experience is the framework where the renders, videos, laboratories, and main events are hosted. For this, the use of Bootstrap was implemented, which is basically an open-source multiplatform library; its handling implies the use of HTML and CSS, which makes it an extremely easy tool to use. There are some free templates which can be manipulated at the user's convenience, in combination with a specific model that includes the kind of

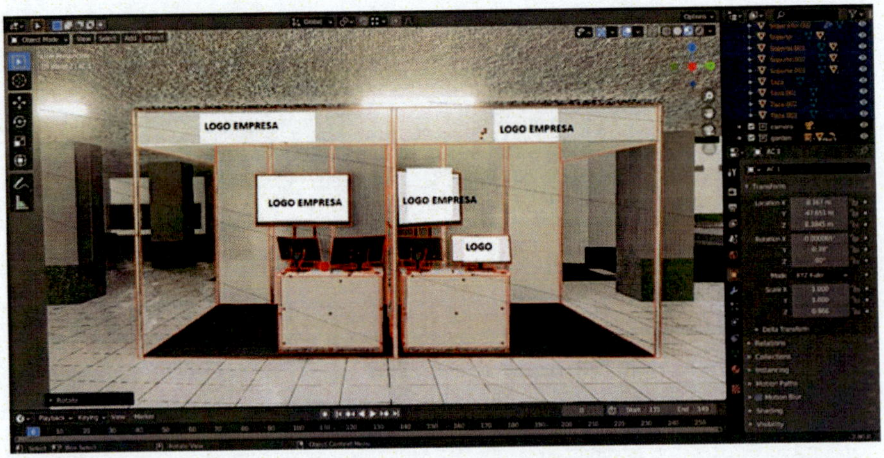

Fig. 4 Stands of the companies made by the 3D department for the implementation of the Virtual Work Fair

Fig. 5 Implementation of the VR experience in Bootstrap

interaction that will be required based on VR games for health interaction research [5] (Fig. 5).

6 Unity 3D

At the time of planning the job fair project, several proposals were made to carry it out; before entering A-Frame, the first option that emerged was Unity 3D because it is a platform with which many collaborators are already familiar and it is an

Fig. 6 Microsoft development platform for 3D projects

Fig. 7 Image of the project in the Unity 3D version

extremely powerful development engine that enables the creation, design, and operation of a fully interactive environment for the user.

One of its main features that makes Unity 3D stand out is that it supports the export of a large number of platforms apart from being supported by operating systems such as Windows, macOS, and Linux Experimental. For the virtual fair, version 2018.3.8 was used, which asks for basic requirements such as Windows 10, Google Chrome, or, where appropriate, Mozilla Firefox and to have a graphic card of at least 1 GB (Figs. 6 and 7).

7 A-Frame vs. Unity

At both platforms, a methodology to know the estimation of effort [6] to build the VR project. When creating virtual reality projects, there are various tools; for the Virtual Job Fair, two were considered, Unity 3D and A-Frame. After a search in the comparison with both tools, the following conclusion was reached: Unity is a powerful tool and A-Frame is a fast tool. Unity is a complete engine to develop games, movies, and virtual experiences as it has been done with the job fair to the point of being at the level of the large video game companies and movie studios. Some advantages of this tool are that it is considered that once you understand the

Fig. 8 Comparison between Unity and A-Frame

jargon and usability, you can create anything. It already contains many premade elements, animations, and motion systems, and if they do not exist, they can be created using C#. Some of the cons are that because it is such a powerful tool, this implies that it is an extremely heavy tool, even making compilations much larger than really necessary, and the processing power to run and load the compilations is not reasonable for the result obtained. For a computer with sufficient resources or a video game console, this may not represent a problem, but when considering it in a mobile environment, the results will not be the most optimal.

On the other hand, A-Frame is a complete, light, fast tool with compilation sizes that will not affect RAM using JavaScript.

The cons of A-Frame is that it is not as flexible as Unity; this is because Unity is a game engine with extensive potential where a combination of script and integrated components is made; on the other hand, A-Frame is a library of JavaScript (Fig. 8).

> A Virtual Reality experience is worth more than a thousand images. (Carlos Fernández (Iberdrola, SA))

8 Conclusions

Currently, there may be thousands of tools that fit our needs [7, 8], some more powerful, some faster and lighter; in the cases that we study, we consider Unity 3D and A-Frame as two main keys for the development of virtual environments; on the side of Unity is the extensive usability, in which if you know how to handle it correctly, you can design any element you imagine; on the other hand, there is A-Frame, a tool perhaps not so powerful but that offers a great virtual experience, great adaptation, and, most importantly, a great compatibility so that practically any user can have an experience in virtual reality without needing teams with large resources. The findings at this approach help to support a next research where the uses of these platforms could be used in other virtual environment experiences as industrial areas [9], or tourism sector [10], to look for their virtual workforce.

9 Remarks

First of all, I want to thank the researcher team, who with their knowledge and support has guided me through each step of my research to achieve the result I was looking for.

In addition to this, we want to thank the work team that made the Virtual Work Fair possible, to Oscar Leal (developer—WebVR) who had great leadership in the project and an excellent implementation of the proposal since with his experience he could take to carry out the development of the Fair in new technologies, to Eduardo Sanchez (vice principal of information technologies/project coordination) who gave us his full support for the design of the project on the Unity 3D platform (without his knowledge, it would not have been possible to carry out an analysis of the differences between the VR design engines), to Mario Gutierrez (FIME 3D Direction) and the FIME 3D department who did an impeccable job in the design and implementation of the 3D environments that with great detail and skill were created, and to Danna Gomez (developer—Web) who was in charge of the implementation of the virtual experience for the end user that very quickly integrated various components to test the project, and we also would like to thank the back end team, Rodrigo Villazuza (database administrator (DBA)) and Valeria Gonzalez (DBA) who were in charge of performing the analysis, development, and implementation of the databases of the project to obtain and store the information for administrative purposes, and finally to Karina Mata who supported the team meetings.

References

1. S.W. Greenwald, A. Kulik, A. Kunert, S. Beck, B. Fröhlich, S. Cobb, S. Parsons, N. Newbutt, C. Gouveia, C. Cook, A. Snyder, S. Payne, J. Holland, S. Buessing, G. Fields, W. Corning, V. Lee, L. Xia, P. Maes, Technology and applications for collaborative learning in virtual reality, in *Making a Difference: Prioritizing Equity and Access in CSCL, 12th International Conference on Computer Supported Collaborative Learning (CSCL) 2017*, ed. by B. K. Smith, M. Borge, E. Mercier, K. Y. Lim, vol. 2, (International Society of the Learning Sciences, Philadelphia, PA, 2017)
2. I.S.A.R. Cosmina, A glance into virtual reality development using unity. Informatica Economica, Acad. Econ. Stud. Bucharest, Romania **22**(3), 14–22 (2018)
3. F. Torres, L.A.N. Tovar, M.C. Egremy, Virtual interactive laboratory applied to high schools programs. Proc. Comput. Sci. **75**, 233–238
4. F. Torres, L.A.N. Tovar, M.S. del Rio, A learning evaluation for an immersive virtual laboratory for technical training applied into a welding workshop. Eurasia J. Math. Sci. Technol. Edu. **13**(2), 521–532
5. L. Neira-Tovar, I. Castilla Rodríguez, A virtual reality tool applied to improve the effects on chronic diseases – case: emotional effects on T2DM, in *Lecture Notes in Computer Science (including subseries Lecture Notes in Artificial Intelligence and Lecture Notes in Bioinformatics)*, vol. 10280, (Springer, 2017), pp. 417–425. https://doi.org/10.1007/978-3-319-57987-0_34

6. F. Ennui, L. Torres-Guerrero., Neira-Tovar, IM. Methodology for the estimation of effort for a project of virtual reality–a case study. *Garcia International Conference on Virtual, Augmented and Mixed Reality*, 82–93
7. E. Kucera, O. Haffner Š. Kozák Kozak. (2018). Virtual tour for smart house developed using Unity 3D engine and connected with microcontroller. Information Technology Applications. VI (2018)
8. D. Souza, P. Dias, B. Sousa Santos, Choosing a Selection Technique for a Virtual Environment, in *Virtual, Augmented and Mixed Reality. Designing and Developing Virtual and Augmented Environments*, VAMR 2014. Lecture Notes in Computer Science, ed. by R. Shumaker, S. Lackey, vol. 8525, (Springer, Cham, 2014). https://doi.org/10.1007/978-3-319-07458-0_21
9. J. Wilson, M. D'Cruz, S. Cobb, R. Eastgate, *Virtual Reality for Industrial Applications Opportunities and Limitations* (Nottingham University Press, Nottingham, 1996) 166 págs
10. Y. Hu, W. Sun, X. Liu, Q. Gan, J. Shi, Tourism demonstration system for large-scale museums based on 3D virtual simulation technology. The Electronic Library **38**(2), 367–381 (2020). https://doi.org/10.1108/EL-08-2019-0185

Towards the Comprehensive Detection of Fake News in Socio-digital Media in Mexico with Machine Learning

Carlos Augusto Jiménez Zarate and Leticia Amalia Neira Tovar

Abstract Contemporary society has made information and communication technologies (ICT) a fundamental axis for socio-digital interaction, we live increasingly hyperconnected through the Internet, and although there are various platforms for socio-digital interaction, only some have achieved globalization and a massive reach of users and have established themselves among the users of social networks on the Internet, to disseminate news content. But the massive use of socio-digital platforms has brought with it harmful practices such as the dissemination of fake news. This research is a review of the literature, of investigations that have used machine learning and other techniques or methods such as natural language processing, text classification and neural networks, for the detection of fake news, to develop an algorithm for the automatic detection of fake news, in the socio-digital content broadcast in Mexican Spanish.

Keywords Socio-digital networks · Fake news detection · Machine learning

1 Introduction

Between 2004 and 2006, Facebook and Twitter were founded; both platforms represented a turning point in social communication. Since then, new platforms for socio-digital interaction have been generated; these new social networks are also known as social media. It is clear that each year more users of social networks are added, and with it the increase in the flow of content and information between people. According to the "15th Study on the Habits of Internet Users in Mexico" [1] presented by the Internet MX Association, it revealed that Internet users in Mexico were 82.7 million and that of these 82% (67.8 million) are users who use the Internet to access social networks and 76% (62.8 million) of users use the Internet to search for information.

C. A. Jiménez Zarate (✉) · L. A. Neira Tovar
Universidad Autónoma de Nuevo León, San Nicolas de Los Garza, Nuevo Leon, Mexico

F. Torres-Guerrero et al. (eds.), *2nd EAI International Conference on Smart Technology*, EAI/Springer Innovations in Communication and Computing,
https://doi.org/10.1007/978-3-031-07670-1_5

Socio-digital interaction has brought with it an increase in negative practices to social communication, such as misinformation and false news or fake news; an investigation carried out by Knight Foundation (2018) revealed that more than one million false tweets are published per day on Twitter. Various civil and government organizations have called this misinformative phenomenon as infodemic; within the context of the COVID-19 pandemic, international health organizations such as the World Health Organization (WHO) and the Pan American Health Organization (PAHO) issued on May 1, 2020, a document to explain the infodemic [2].

The fight against fake news has taken on greater relevance; due to the COVID-19 pandemic, on May 21, 2020, the UN issued a tweet from its official account showing the launch of a tool to verify news through the portal "Shareverified.com" under the motto "There has never been a greater need for accurate and verified information."

On Twitter, content dissemination campaigns have been generated or developed with negative aspects, such as the spread of disinformation, fake news, the use of bots or automated digital positioning systems, or the dissemination of negative or hateful content, as indicated by Stella et al. [3].

Mønsted et al. [4] determined that socio-digital networks generate structures that propagate content or information, which can be analyzed with greater efficiency through complex contagion models, and also assume that the probability of content adoption depends on the number of unique sources of information. In this context, the complex contagion of content propagation bots can be targeted as news disseminators with erroneous information, disinformation, or fake news.

In the "Report on disinformation campaigns, fake news and their impact on the right to freedom of expression," which was prepared by the National Human Rights Commission of Mexico [5], warns about the use of content with information that is not attached to objective or inaccurate facts, and that exploits the emotions or beliefs of the audiences, to attract more "likes" or "retweets" on the Facebook and Twitter platforms; it also mentions that most of the citizens do not have the time, resources, or instruments to verify the content or information they receive in an increasingly connected society.

In the bulletin UNAM-DGCS-318, the study entitled "Radiography on the Dissemination of Fake News in Mexico" is mentioned [6], prepared by UNAM, which ensures that at least 89% of users on Twitter in Mexico have been exposed to this type of content. Fake news affects various areas, such as business marketing, as determined by Visentin et al. [7], who investigated the negative effect on brands that advertise in media or sites that spread fake news; their results indicate that marketing managers should be encouraged to monitor, since the proliferation of fake news, constitutes there is an increasing risk for the business sector. The increase in the use of socio-digital platforms has brought with it an increase in harmful practices to social communication, such as the spread of fake news, the use of bots, trolling, and artificial positioning strategies. Vosoughi et al. [8] determined that fake news spreads further, faster, and deeper than real news on social networks like Twitter, as well as fake news on political issues; they have a higher level of spread than other social topics.

For the analysis of this growing and continuous amount of digital information, it is necessary to implement machine learning algorithms, as well as other artificial intelligence tools, because these tools provide a great capacity for data processing and have become an indispensable support for the development of highly competitive predictive models.

2 Literature Review

To guide the investigation on the construction of a system for the detection of fake news on Twitter in Mexico, this investigation has condensed the most relevant investigations that can help its development. Shao et al. [9] analyzed 14 million tweets, where 400 thousand articles were shared during 10 months between 2016 and 2017; through their analysis, they managed to find evidence that much of the misinformation is due to super propagators that are social bots that publish automatically links to articles; the analysis tools they used were the Hoaxy and Botometer verification systems, which were developed by researchers at Indiana University. Davis et al. [10], developers of the BotOrNot system, claim that the classifier generates more than 1000 characteristics through the use of metadata and information extracted from patterns and the content of the interaction.

The Hoaxy system collects public tweets that contain links to news; the platform is freely accessible and allows systematic studies on a large scale, on topics or hashtags that are part of a fake news dissemination strategy. Shao et al. [11] used the Hoaxy platform to carry out an investigation of the dissemination of erroneous information before and after the US presidential election in 2016, the study was based on the analysis of the core of the propagation networks, determined that the network of users is polarized between true or false information. The dissemination and propagation of fake news cover different areas or themes of society, but they can also be categorized by dimensions as described by Shu et al. [12], who determined three types of dimensions (content, social, and temporal); their research made it clear that fake news is not an insignificant matter, since they are built to lie to readers and propose from an analysis point of view of social networks, a method of inoculation before the spread of fake news, which consists of identifying the nodes, routes, or main propagation links, and with this information, strategies for inoculation, blocking, or containment can be created. Ahmed et al. [13] focused on the detection of spam and fake news through text classification, for which they developed a new n-gram model.

The detection of fake news is not easy; it requires models and systems that can summarize and compare the news with reliable sources to be able to categorize them; that is why alternatives are sought such as identifying the position through the automatic detection of the relationship between two pieces of a text. Thota et al. [14] developed a model where they used the deep learning architecture of neural networks, with vectorization through a bag of words with a dense neural network, to be able to categorize the positions; the model showed good results to categorize the headings and new articles or news. Altunbey et al. [15] compared more than 20 supervised artificial intelligence algorithms for the classification of fake news and determined that the decision tree algorithm obtained a better result.

Oehmichen et al. [16] determined the characteristics of the accounts that spread fake news; in their research, they started from the creation of a dataset, for which they collected for 4 months the tweets related to the hashtags of the presidential election in the USA of 2016; they took into account tweets greater than 1000 retweets and managed to create a dataset of 9001 tweets. They determined that the fake news spreading accounts are recently created; the vast majority are unverified, have fewer updates, use strange characters in the name and description, have few followers and follow many more, and are generally dedicated to interact with retweets.

Fake news can be used to stifle social protest movements. Zervopoulos et al. [17] used various machine learning techniques, such as naive Bayes, support vector machine, C4.5, and random forest, to be able to classify the characteristics linguistics of the fake news; for this, they took the tweets in English and Chinese from a Twitter database to be able to classify the fake news. Zhou et al. [18] proposed a multimodal analysis system, which integrates the textual and visual analysis of the news; for this, they resorted to the construction of a dataset with information from news verification sites in the USA.

An important space for analysis in the Spanish language has been the development of the Semantic Analysis Workshop (TASS) which is part of the actions of the Spanish Society for Natural Language Processing (SPNL), which aims to encourage semantic analysis in Spanish language. In this effort, the IBERLEF (Iberian Languages Evaluation Forum) has been integrated, where a competition is developed to encourage research for word processing for Iberian languages such as Spanish, Portuguese, Catalan, Basque, and Galician. Salas et al. [19] implemented an analysis scheme, using a system of machine learning algorithms, a model to determine Spanish and Mexican satire on Twitter. The results of their research showed a high accuracy for detecting satire and that there is no significant difference in satire from both countries.

Posadas et al. [20] conducted an investigation to detect fake news in the Spanish language, for which they created a new dataset of the content broadcast on Twitter by formal media and media that regularly publish false content; they used four algorithms for classification machine learning, which were support vector machine, logistic regression, random forest, and boosting. Within this review, very few investigations focused on the detection of fake news for the Mexican Spanish language were found.

Table 1 summarizes the investigations that are considered relevant due to their methods for the construction of an efficient system for the detection of fake news in Twitter Mexico networks.

Table 1 Literature review. Most relevant papers for detecting fake news with machine learning

Author(s)	Research title	Method
Shao et al.	Hoaxy: A Platform for Tracking Online Misinformation (2016)	Extraction of tweets containing the URLs of the websites
Shao et al.	The Spread of Low-Credibility Content by Social Bots (2017)	Extraction of articles shared on Twitter, verification of users who shared the articles or content of low credibility
Ahmed et al.	Detecting Opinion Spams and Fake News Using Text Classification (2017)	They used 3 existing datasets and created a new one made up of 12,600 fake news and 12,600 legitimate news. In addition, they used two extraction techniques and six learning classification techniques (stochastic gradient descent, linear support vector machines, K-nearest neighbor, logistic regression, and decision trees)
Salas et al.	Automatic Detection of Satire in Twitter: A Psycholinguistic-Based Approach (2017)	They created a dataset with satirical and non-satirical news and obtained data from Mexican and Spanish sites. And for the analysis they used machine learning algorithms for their classification
Vosoughi et al.	The Spread of True and False News Online (2018)	They created a dataset with 500K reply tweets that included the links from the verification sites to other tweets (original tweet). Later, from the original tweet, I determine its cascades of propagation. This in order to quantify the spread of rumors or false news
Thota et al.	Fake News Detection: A Deep Learning Approach (2018)	They used three different neural network architectures, with the TF-ID vector representation of combined words being the best performing one
Altunbey et al.	Fake News Detection Within Online Social Media Using Supervised Artificial Intelligence Algorithms (2019)	They developed a two-phase method. The first is to convert unstructured data into structured data, using weighting vectors. The second phase is an experimental evaluation of 23 supervised algorithms
Oehmichen et al.	Not All Lies Are Equal. A Study into the Engineering of Political Misinformation in the 2016 US Presidential Election (2019)	They created a dataset of 9001 tweets with more than 100 retweets, referring to the 2016 presidential election in the USA. They used syntax and sentiment analysis
Posadas et al.	Detection of Fake News in a New Corpus for the Spanish Language (2019)	They created a news dataset of formal media and sites that regularly post fake news. They used machine learning algorithms such as support vector machine, logistic regression, random forest, and boosting
Zaizar et al.	ITCG's Participation at MEX-A3T 2020: Aggressive Identification and Fake News Detection Based on Textual Features for Mexican Spanish (2020)	They used the dataset from the MEX-A3T Natural Language Processing Assessment Forum, and for classification they used the Natural Language Toolkit (NLTK) for tokenization and stopword removal. For the training of their model, they made use of cross validation with learning algorithms
Zhou et al.	SAFE: Similarity-Aware Multi-modal Fake News Detection (2020)	They used a multimodal approach that integrates the visual textual analysis of the news, to be able to categorize them as true or false; the dataset was obtained from news verification sites in the USA

3 Proposed Method

The monitoring and analysis of socio-digital interaction have become essential for the analysis and planning of communication strategies and socio-digital interaction, either to know social opinion, or in the development of strategies for digital marketing campaigns in business sectors, social or political, as Antoniadis et al. [21].

This research proposes the implementation of machine learning algorithms for data processing and analysis, since they are capable of systematically analyzing large dataset and being able to categorize them without the interference of human bias.

After reviewing the literature and the state of the art regarding the detection of fake news on Twitter Mexico, this research proposes to use text categorization techniques for the body or text of the news (URL in tweet) indexed in the tweet, for the classification of users using the random forest algorithm. For the analysis of the spread of content, it will be done using the concepts of social network analysis and the method to determine diffusion cascades of Goel et al. [22].

The independent variables for the development of this research will be:

(a) Tweet text
(b) Text of the news (URL in tweet)
(c) Broadcast users
(d) Propagation of the tweet

The dependent variables will be the following for the text of the tweet and the text of the news that is contained in the URL of the tweet:

(a) Fake news
(b) Satire
(c) Propaganda
(d) Real news

The dependent variables for the independent variable of @user will be:

(a) Bot
(b) Troll
(c) Human

The dependent variables for tweet propagation will be:

(a) Viral
(b) Non-viral

The independent and dependent variables proposed will be integrated into a system that may be capable of detecting false news and other variants of the news content broadcast on Twitter Mexico (Fig. 1).

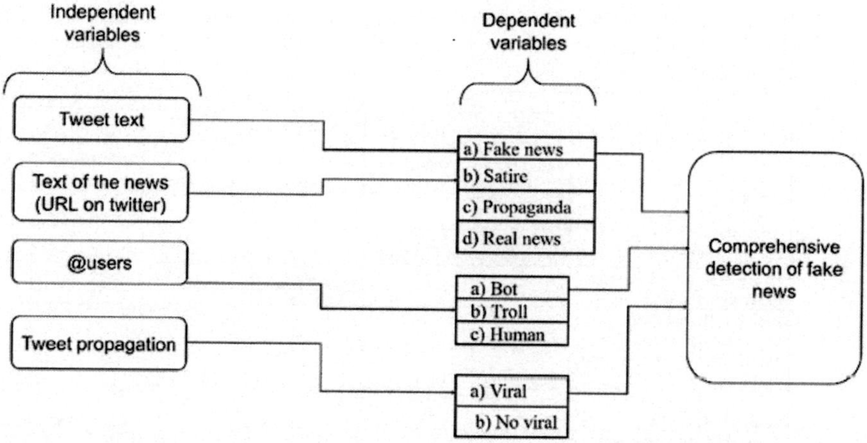

Fig. 1 Block diagram of the independent and dependent variables for the model proposed for the detection of fake news in the socio-digital media in Mexico

4 Development Phases

The first phase of the proposed system for detecting fake news on Twitter Mexico with machine learning will be the review of the literature on the various automatic techniques for detecting fake news.

The second phase consists of creating a dataset with tweets in Mexican Spanish for training and testing. For this, you will need to partition the dataset into two datasets:

- The first training partition will be by manual classification of the tweets and is composed of 70% for the dataset.
- The second partition will contain 30% of the dataset for the test dataset.

The training dataset will be processed and analyzed with various machine learning word processing algorithms. With the analysis of the text, the tweet can be classified to determine if it is fake news, satire, propaganda, or real news.

The third phase will consist of designing and building a comprehensive algorithm for the detection of fake news in Mexican Spanish. That you will have to analyze and categorize fake news automatically and massively.

The fourth phase consists of testing the algorithm and analyzing the results and verifying if the proposed algorithm has an acceptable degree of efficiency in execution time and of efficiency in the detection of fake news.

The fifth phase will consist of communicating the results in a completed research paper.

In Table 2, you can see the phases to carry out the project of the algorithm for the detection of fake news in the socio-digital networks in Mexico. Currently, the project is in the phase of building the dataset for further training and testing. The later stage is the design and construction of the algorithm for the detection of fake news.

Table 2 Project calendar for the construction of the algorithm for the detection of fake news in the socio-digital networks in Mexico. The boxes in green are the actions that have already been carried out

Phases	2020		2021		2022	
	Time Interval I: January to June	Time Interval II: August to December	Time Interval III: January to June	Time Interval IV: August to December	Time Interval V: January to June	Time Interval VI: August to December
Phase 1: Requirements Analysis and Existing Investigations						
Phase 2: Build the dataset for training and testing						
Phase 3: Design and construction of the algorithm for the detection of fake news						
Phase 4: Testing and analysis of results						
Phase 5: Communication of results						

5 Conclusion

The literature review found that studies and research on the detection of fake news with machine learning or artificial intelligence techniques have been carried out mainly in the English language or with translators. The most used techniques for the classification and detection of fake news have been those of SVM, logistic regression, decision tree, and naive Bayes.

During the literature review, it was observed that throughout various investigations, various datasets have been created that have ranged from official news media, alternative media, journalists' accounts, as well as fake news verification websites and also fake news websites and satirical content. The topics that have covered the papers read have been mainly on politics and society.

For the Mexican Spanish language on Twitter, there is very little research that has been carried out in recent years, and they are limited to analyzing the text of the news that is attached through a link or link in a tweet; this limits the scope for detection, since it leaves out add-ons that can be integrated for a better detection of fake news.

The proposed research will be useful in the first place for the detection of false news that is disseminated on social networks, and it will also be a tool that helps to report content detected as false news in the different socio-digital networks. The system can also be of great help in the development and monitoring of social communication strategies for any type of organization, be it business, social, governmental, or political. The use of this algorithm will be useful to improve the veracity of the contents in the socio-digital networks.

References

1. Movilidad en el Usuario de Internet Mexicano. 1–25.
2. E.S. Decisivo, G. Scholar, Entender la infodemia y la desinformación en la lucha contra la COVID-19. **395** (2020)
3. M. Stella, E. Ferrara, M.D. Domenico, Los bots aumentan la exposición a contenido negativo e inflamatorio en los sistemas sociales en línea. **1–21** (2020)
4. B. Mønsted, P. Sapieżyński, E. Ferrara, S. Lehmann, Evidence of complex contagion of information in social media: an experiment using Twitter bots. PLoS One **12**, 1–14 (2017)
5. C. Itzel, P. Farfán, Reporte sobre las campañas de desinformación, “no ticias falsas (fake news)” y su impacto en el derecho a la libertad CNDH. México (2019)
6. UNAM-DGCS-318. Radiografia de la propagacion de fake news en México. 1–2 (2020)
7. M. Visentin, G. Pizzi, M. Pichierri, ScienceDirect fake news, real problems for brands: the impact of content truthfulness and source credibility on consumers' Behavioral intentions toward the advertised brands. J. Interact. Mark. **45**, 99–112 (2019)
8. S. Vosoughi, D. Roy, S. Aral, News On-line. Science (80-.). **1151**, 1146–1151 (2018)
9. C. Shao, G.L. Ciampaglia, A. Flammini, F. Menczer, The spread of low-credibility content by social bots. Nat. Commun. (2017). https://doi.org/10.1038/s41467-018-06930-7
10. C.A. Davis, O. Varol, E. Ferrara, A. Flammini, F. Menczer, BotOrNot: A System to Evaluate Social Bots. 4–5 (2016). https://doi.org/10.1145/2872518.2889302
11. C. Shao, G.L. Ciampaglia, A. Flammini, F. Menczer, Hoaxy: A Platform for Tracking Online Misinformation (2016). https://doi.org/10.1145/2872518.2890098
12. K. Shu, H.R. Bernard, H. Liu, Studying Fake News via Network Analysis: Detection and Mitigation. 43–65 (2019). https://doi.org/10.1007/978-3-319-94105-9_3
13. S. Ahmed, K. Hinkelmann, F. Corradini, Combining machine learning with knowledge engineering to detect fake news in social networks – a survey. CEUR Workshop Proc. **2350** (2019)
14. A. Thota et al., Fake news detection: a deep learning approach. SMU Data Sci. Rev. **1**, 10 (2018)
15. B. Altunbey, Fake news detection within online social media using supervised artificial intelligence algorithms. Phys. A Stat. Mech. Appl. **540**, 123174 (2020)
16. A. Oehmichen et al., Not all lies are equal. A study into the engineering of political misinformation in the 2016 US presidential election. IEEE Access **7**, 126305–126314 (2019)
17. A. Zervopoulos et al., Hong Kong protests: using natural language processing for fake news detection on twitter. IFIP Adv. Inf. Commun. Technol. **584 IFIP**, 408–419 (2020)
18. Zhou. Syracuse: Detección multimodal de noticias falsas con reconocimiento de similitudes Abstracto. 1–21 (2020)
19. M.d.P. Salas-Zárate, M.A. Paredes-Valverde, M.Á. Rodriguez-García, R. Valencia-García, G. Alor-Hernández, Automatic detection of satire in Twitter: A psycholinguistic-based approach. Knowledge-Based Syst. **128**, 20–33 (2017)
20. J.P. Posadas-Durán, H. Gomez-Adorno, G. Sidorov, J.J.M. Escobar, Detection of fake news in a new corpus for the Spanish language. J. Intell. Fuzzy Syst. **36**, 4868–4876 (2019)
21. I. Antoniadis, P. Serdaris, A. Charmantzi, The application of social networking analysis in marketing: a case study of a product ' s page in Facebook. *2nd Int. Conf. Contemp. Mark. Issues* (2014)
22. S. Goel, A. Anderson, J. Hofman, D.J. Watts, The structural virality of online diffusion. Manage. Sci. **62**, 150722112809007 (2015)

Virtual Reality Training Simulation for Controlled Land Unmanned Vehicle: A Feasibility Model

Lozano González Jorge, Neira-Tovar Leticia, Cruz Juan, and Sanchez Eduardo

Abstract Today, the use of technology is indispensable in daily life, being this an important part to facilitate to the greatest extent possible the activities carried out by human beings as an example of the scope that technology has and the impact it generates in our lives. Having its origins in the last century, virtual reality has become one of the most sophisticated systems with different applications in different environments. This chapter presents an overview of a virtual reality (VR)-based training tool for users to get qualified as pilots capable of handling real-life unmanned vehicles. This work shows some characteristics of VR systems, and the scope is to present the advantages of using virtual reality software as a training tool, promoting that virtual reality training for use of an unmanned vehicle could be as effective as real-life training.

Keywords Virtual reality · Augmented reality · Simulation · Unmanned vehicle · Training software

1 Introduction

Pioneering organizations such as GE and Boeing are using virtual reality (VR) to improve productivity, quality, and training [1]. By combining the power of humans and machines, VR will significantly increase value creation for organizations [2]. VR is a very high-end computer interface that evolves real-time simulation and interface through numerous sensorial channels. These sensorial modalities are visual, aural, tangible, smell, taste, and other senses [3]. On its part, augmented reality (AR) is a new technology that involves the overlay of computer graphics on the real world [4]. They both aim to extend the sensorial environment of an individual by mediating reality through technology. The former relies on an

L. G. Jorge (✉) · N.-T. Leticia · C. Juan · S. Eduardo
Universidad Autónoma de Nuevo León, San Nicolás de los Garza, Nuevo León, México
e-mail: jorge.lozanogz@uanl.edu.mx

F. Torres-Guerrero et al. (eds.), *2nd EAI International Conference on Smart Technology*, EAI/Springer Innovations in Communication and Computing,
https://doi.org/10.1007/978-3-031-07670-1_6

alternative setting to experience, while the latter improves existent elements with additional layers of meaning [5]. As we can see, these have many similarities, but differences as well.

In this work, we are going to focus on VR systems; VR is known as a technology that makes possible for people an experience in a world beyond reality. Such technology comes in different ways and sizes, with the aim of giving information to our senses: sight, hearing, and touch [6]. This technology qualifies better for this project thanks to its features. VR systems immerse the user in a virtual world that can be created according to the characteristics we want it to have.

Although the use of virtual reality is, in general, highly related to the world of video games and entertainment, we cannot ignore the applications that this may have, such as the formation of new employees in different industry sectors; within medicine, the most representative applications are simulators for medical training, surgery operations, treatment of phobias and psychological trauma through exposure therapies and pain management through distraction techniques, and others. In addition to providing technology with the possibility of disseminating knowledge and interacting with it, once each training program is invested, the expenditure on employee training will also be reduced. As we mentioned, all of these provide employees with a more attractive and effective environment, giving them more motivation, interactivity, or connectivity without any risk for the user or monetary losses for the company.

Based on the above, the project seeks to implement a video game-type simulator that tests the skills of users who test the simulator, thus improving their ability to control an unmanned land vehicle to perform different activities with real prototypes without additional costs or material damage that could cause delays in the processes in where they are used. To achieve this, it is recommended that the simulator has an intuitive interface that any user can control to make your experience more satisfactory. In addition, some levels have been developed so that according to the standards of user experience, the user can choose the level of difficulty he needs to improve his skills or, on the contrary, gradually increase the level and get a better training.

Virtual reality devices are expected to grow rapidly over the next 5 years, with an average CAGR of 106% and combined shipments of 43 million in 2020. Augmented reality smart glasses will experience similar growth and ship 21 million units in 2020 with a CAGR of 78%. "Sales of VR Head Mounted Displays, including mobile-based devices like the Samsung Gear VR and Google Cardboard, as well as tethered devices like the Oculus Rift and Sony's Morpheus, will be driven by the release of high-profile devices and a growing awareness and interest in the technology from consumers" [7]. Due to the above, it is expected that the project will be a reliable and effective tool for companies in different sectors in which a prototype of both the simulator and the unmanned vehicle can be useful.

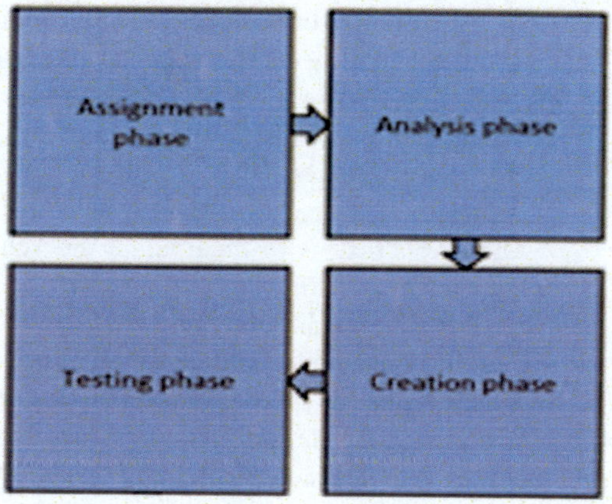

Fig. 1 Phases of methodology

2 Method

Virtual reality (VR) uses simulation and displays to trick humans and other organisms into believing they are having a perceptual experience that is different from reality. This experience is usually interactive and carefully crafted, such as games exploring exotic worlds [8]. The effectiveness of the VR-based training process depends directly on the quality of the prepared training material or on the quality of the prepared virtual training environment. A detailed overview of the product documentation and the process of product assembly (installation maps, design drawings, 3D models) enable the definition of all the elements necessary for defining the training process [9].

The method employed is based on a "Methodology for Designing Virtual Reality Applications," after rigorous evaluating sessions with people who had been involved in previous virtual reality projects and analysis of other methodologies mentioned inside the document as well. After deployment on the first projects, they made some corrections, and this methodology was further enhanced [10]. This work is based on four main steps (Fig. 1).

2.1 *Assignment Phase*

The fundamental elements that determine the impact that any education technology tool can make are informed and determined not only by its cost or speed or accessibility but also by its usability for a diverse and inclusive society [10]; therefore, in this phase, we check the main ideas of the overall concept of the virtual reality application we are going to develop; we must be able to describe the requirements and purpose of our project. Basically, feedback was given on what we wanted to achieve with the release of the project.

First of all, a brainstorm was done to have different options on how to implement a project that meets the expectations but that mainly fits within the time that we had to be fulfilled. There were many complications about how the project would be carried out, as there were many ways to achieve it. However, not all of them fit the characteristics that were mentioned; therefore, taking ideas from UAV's flight simulators, it was decided to develop the simulator as a video game, in which the user's task is to manipulate the vehicle by making it go through an obstacle course, thus picking up the objectives, being this one the best option for development process due its difficulty and best training effectiveness.

For convenience and better training, the simulator was designed to contain ten levels, each one more difficult than the previous ones. Each level was designed with the same purpose, but the easier ones were designed to learn most basic movements of the vehicle, such as directing on a plain with simple obstacles, while in the harder ones, we can see an increase in the difficulty like having alterations as slopes and holes, making it more difficult for the driver, because it no longer has to face these changes, but requires precise control of the vehicle to avoid falling over the edges and other obstacles.

2.1.1 2D Unity Requirements for the Simulator

1. Creation of a simple main interface with the name of the simulator and with noncomplex navigation. The important thing is to capture the user's attention by not being so invasive.
2. Creation of the submenus corresponding to each button. The design of all of them remains similar to the main menu but with their respective title and different options in each one.

2.1.2 CAD Requirements for the Simulator

1. Build a virtual model of a prototype unmanned land vehicle that will be used as our object to control throughout the tests of the simulator.
2. To design levels, which will gradually increase in difficulty, requiring the user to have different skills and making him go through different situations through which he will have to go through to complete the different levels and conclude if the user is able to control a prototype in real life.
3. In each level, there will be objectives in the form of medals, which will have to be collected to complete it. The design of these medals will also be carried out for their subsequent placement in the simulator. Since the levels will be surpassed once the user manages to collect all the medals, it is important to consider a design that is striking and easy to locate for users.
4. The construction of a pad, in which the user must be placed to later move to the next level.

2.1.3 Unity Editor Requirements

1. Import the models in .FBX format and add them to the "Models" folder.
2. Development of scripts for the physics of the simulator.
3. Placing the objects and respective textures to each one.
4. Edition of the cameras and lighting.
5. Programming the actions of the objects, as well as the physics ones of each one.

2.2 *Analysis Phase*

In this section, we analyzed what would be the necessary requirements for the creation of the simulator: the selection of actions and objects and designs that will be placed in the final program. Processes establish how methods are applied and software projects are managed. Methods supply technical descriptions on how to perform different tasks of software engineering, such as designing and building software. Tools include computer programs used for planning and implementing software, as well as otherwise supporting the process [11]. In this phase of software development, the goals of Software Safety Program are to eliminate, reduce, or control the possible hazards associated with potential software failures. Software safety requirements may include national/international standards, customer requirements, or corporate needs [12].

The first thing that had to be thought about was the menu design. The menu shows the logo and the name of the simulator at the top; the name LUV-SIM originated from the combination of the initials of land unmanned vehicle with the word simulation. Since this is the presentation of the simulator and it must achieve a positive impact on the user, using a simple and friendly interface in which he can make the adjustments he thinks is necessary, in addition to having the option of selecting the level of his preference into the play menu.

Regarding the structure of the simulator, the main thing was to choose how the levels would be structured and which skills would be required in each of the proposed levels and how they would be distributed to achieve adequate training. The objective at each level is simple, to collect the floating coins found along the map and then place the vehicle on a podium that will shine exclusively when the user has the coins found at each level. Levels were designed so that the user can first learn the basics, such as the basic movements of the vehicle, as well as get the user to adapt to the controls so that with the passing of the levels and already with some adaptation, he can successfully conclude each of the challenges presented.

As for the structure of the vehicle, it is an original model that is under development but serves, for the moment, as a representation of the prototype used exclusively for this simulator. The most significant details of the model are the adaptation of cameras and LEDs that the pilot of the prototype can use according to their needs (Fig. 2).

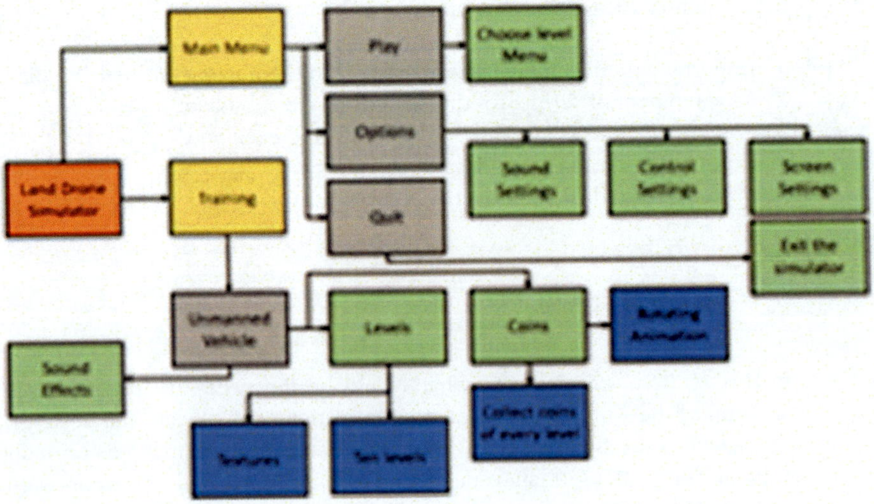

Fig. 2 Diagram of analysis phase

The next step is to consider the actions that each of the models will perform within the simulator environment so that the system meets its objective:

1. The vehicle must go in the desired direction, either forward, left, or right, and in reverse.
2. The collision property will be added to each one of the objects that require it; these will be those whose function is to mark a limit or must be immovable for training purposes.
3. The coins should be rotating to take advantage of their texture and be located more easily for users.
4. Each time that each of the located medals is collected, it should disappear both on the map and in the simulator objectives count.
5. If the user has not collected all the medals on the map, pad must not allow him to complete the level.
6. Wheels of the vehicle must move whenever the vehicle is directed in any direction.

2.3 *Creation Phase*

The simulated environment can be like the real world to create a lifelike experience, for example, in simulations for pilot or combat training or it can differ significantly from reality, such as in VR games [13]. Because of this, several factors must be considered when creating a virtual reality environment. The first thing was to think about how real the simulator should look, to make more dynamic the way in which

Fig. 3 Land unmanned vehicle model isometric view

the user develops and aesthetics; it was decided to be a more unreal environment and look more like a futuristic video game.

The development of the project was carried out in different phases and software, of which the first was to develop the structure of the unmanned vehicle that will be used for the tests in virtual reality.

The vehicle model and its parts were made in SolidWorks because of its ease of use and later the assembly was made. Levels were made in SolidWorks to design both walls and floors for each level with no materials selected, to be assigned later in another software.

2.3.1 CAD Design

The measures that were given to the vehicle were assigned considering an acceptable size for the vehicle to meet the applications that would be given (rescue, transport, exploration, etc.). The hitbox dimensions given for the side wheels were 172 mm, being the total height of the vehicle, besides being the largest measure of all. The width of the vehicle is 314.72 mm, and finally, the depth of the vehicle is a total of 210.64 mm. However, the total volume of the vehicle is 772269.89 mm^3 (Fig. 3).

The complexity of the design is justified by the fact that the physical prototype can carry out tasks in which human intervention is difficult and requires human intervention and that it has the tools to carry out the task. It consists of a model that has two wheels at the sides and one at the back to maintain balance; the front is shown with a camera, as well as a night vision sensor and LEDs to facilitate vision as shown in the figures below.

On the other hand, the design of the levels was carried out by means of the difficulty that these would suppose for the user. Each level was planned to require the user in different facets or challenges that must be faced. The difficulty varies depending on the level, with number one being the easiest and the level with the highest number being the most difficult. Throughout the ten levels, users are tested so that the right challenges are demanded from the user and better training is completed. General rules were maintained for the creation of the levels. The walls of all levels have a height of 3000 mm so that in case of a bug or glitch in the simulator, do not leave the assigned play area. Also, the walls are 100 mm thick and the floor is

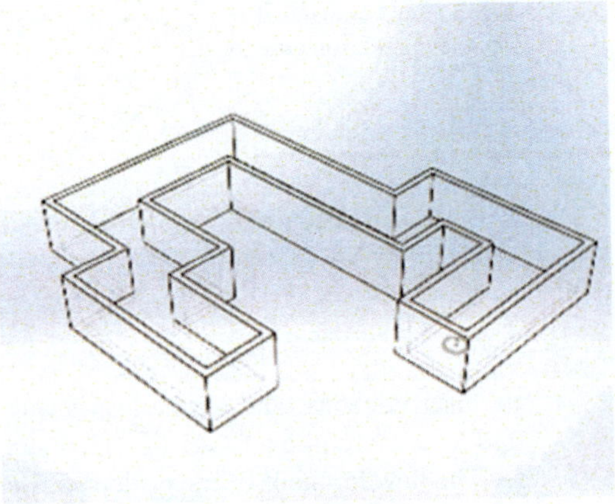

Fig. 4 First level isometric view (hidden lines visible)

Fig. 5 Pad final assembly (no color or textures given)

750 mm wide. All these measures were assigned to each of the levels regardless of the difficulty, to achieve a standard and not have variations between them (Fig. 4).

Finally, the design of the pad in which the vehicle must be placed at the end of each test was constructed. This pad consists of an assembly also designed in SolidWorks. The design is a circle with a diameter of 300 mm and a height of 14.66 mm, the latter in order not to be difficult to climb for the vehicle but nevertheless be prominent from the rest of the environment (Fig. 5).

2.3.2 Blender

Once all the models needed were made, both levels and the model of the vehicle, we had the need to export them to the unity software. For this, the first thing we had to do was to export each of the models in .STL format to be able to modify scales, rotate angles, and convert them to .FBX format for export to the Unity editor where the final project will be done. The image below shows the model of the vehicle in blender for later export (Fig. 6).

Once the models are processed to the corresponding format, we can use the models created in the Unity editor to add textures, place the objects in the corresponding place, and continue with the next step which is the creation of scripts, the addition of physics, and final programming.

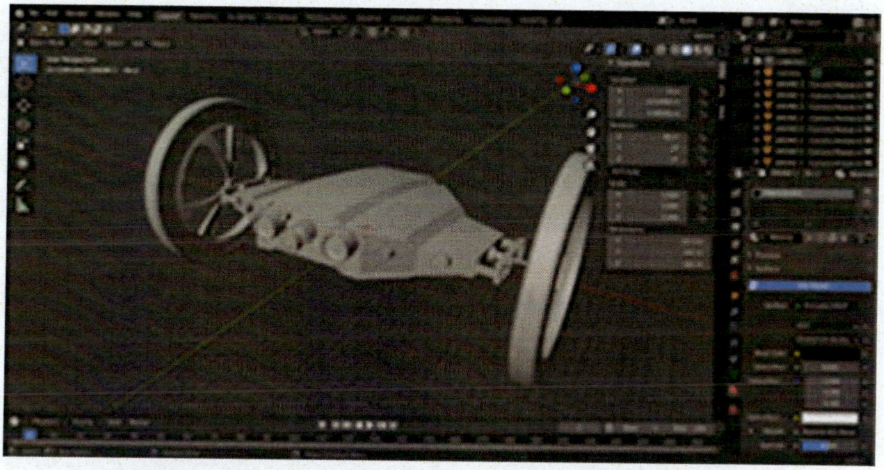

Fig. 6 Unmanned vehicle model in blender

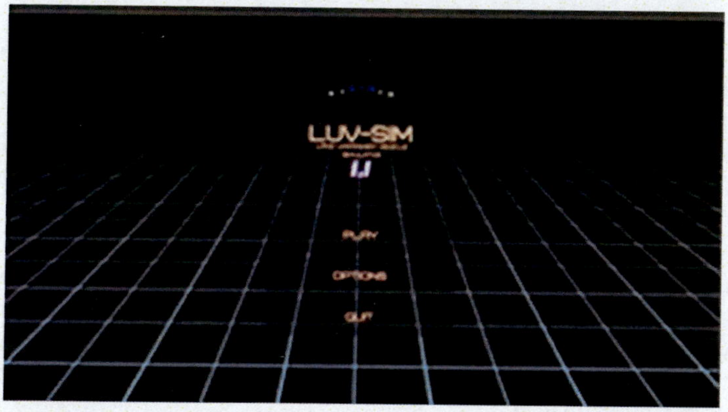

Fig. 7 LUV-SIM main menu

2.3.3 Unity Editor 2D

The first part, already concentrated only in the Unity editor, was the creation of the menu in a 2D project in which the logo was placed at the top of the screen with only three buttons so as not to make the user interface so invasive (Fig. 7).

The design selected for the background kept the idea of a futuristic design so the background being the first selection to be made had to convey this idea, and the grid was selected with a blur, imitating a little the look of the levels inside the simulator. We added the background by right-clicking on the scene where we will place our background, then select the "UI" tab where "Image" was selected to then select the parameters that fit what we wanted to achieve, such as resolution, size, etc.

Fig. 8 LUV-SIM 1.1 logo

Later, we selected the logo. The logo was created in the Canva web software. This design consists of an image of the vehicle with the title already mentioned, "LUV-SIM" adding its version (Fig. 8). This was added just like the background but with different parameters that were selected specifically for the logo, for example, the size and the coordinates where it will be placed. The buttons were added with the "Button Text Mesh" option in the "UI" menu.

By default, the Unity editor adds text to this type of buttons, but it was replaced to have a more uniform structure for the menu and respecting the same font and colors that were given with the logo, since all this is shown in the same scene. Due to this, we exchanged the text for a centered image that shows uniformity in the menu, since it corresponds to the typography of the title and respects the aesthetics of the futuristic idea.

Since now, an image was shown and not text on the buttons, we had to adjust some parameters such as the size of the buttons and find a way they could fit, since the size of the image and the location had to be modified because they were different. Also, it was necessary to select the positions that each button would occupy, so it was decided that the option of placing them in the form of a vertical line would be the most efficient way to not occupy much space on the screen.

However, the simplicity of the menu does not mean that no more options can be placed, since each of the buttons opens submenus in which, for example, the "PLAY" button will open a screen in which the desired level can be selected (Fig. 9).

This section, as its name says, gives us the option to choose the level we want to select, according to the ability of the user. For this, we use the same methodology as in the main menu. We kept the background and the typography handled, being monotonous for all the menus of the simulator and a logo in the upper part, with the difference that in the upper part would be placed the name of the menu in which we

Fig. 9 Choose level menu

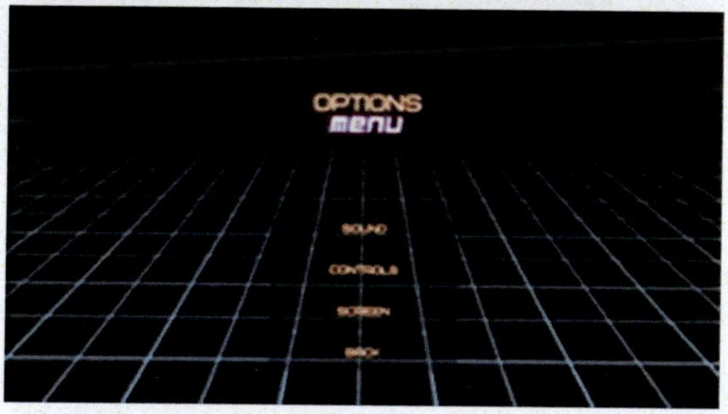

Fig. 10 Options menu

were, in this case the level selection. Each of the buttons maintains the size of the main menu but with a different distribution so that the ten level selection buttons and the button that would take us back to the main menu would fit. In terms of style, the typography was kept as mentioned above, but with the addition that we now had numbers, so it was decided that we would change the color of these to achieve a different aesthetic without losing the essence that was built with the main menu.

Also, the "OPTIONS" button will open a menu in which you can adjust different parameters such as volume, screen settings, and controls (Fig. 10). Keeping the style already proposed in the previous menus, the buttons were placed in a similar way to the main menu with the addition of a back button to the main menu.

And finally, we will have the "QUIT" button that will serve as the output of the program.

2.3.4 Unity Editor

The Unity editor part is the part where the most work was done, due to the development of the scripts and the programming done.

The first part of the process was to import all the models that were designed to be placed in the scenes, which will later become the levels of the simulator. The 3D Models were placed in a folder called "Models" in Unity that was created to put there everything necessary for the creation of each level; this includes the level

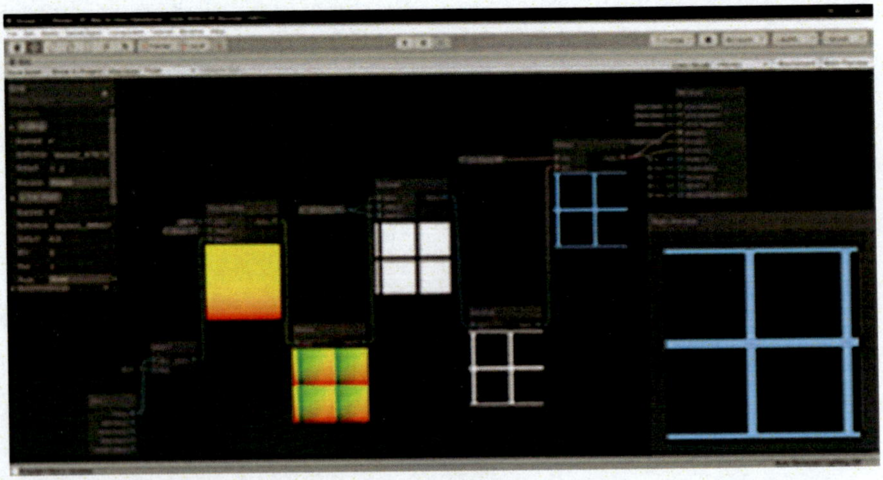

Fig. 11 Grid properties and parameters

itself, the medals that will be placed to complete the level, and if necessary, we must import the pad itself, because although in most levels is already included in the final assembly made before, in case of an error must be placed again and being a completely independent part of the walls and floor, we must return to place it manually. The explanation of the textures will be given by taking the example of a level.

The addition of the grid texture for the walls and floors was done from an animation made first by placing a square with the command rectangle and giving it fraction properties making it multiply along all the structures. Later, the tiling and offset parameters were added with a timer and multiply to generate the motion animation. The connection of these parameters is shown in the figure below (Fig. 11).

Since the vehicle textures had already been added previously in blender, we do not need to place another one for this one. The only ones that were added were to the missing structures that are the pad and the medals. These were not given a dynamic texture since they are moving objects. We just had to add a solid texture that would fit the look of the simulator. The medals, being the main objective of the simulator, were given a golden texture that could be easily seen by the user (Fig. 12).

Once the textures were finished, the scripts were configured with the physics needed to make the simulator as similar as possible to a model built in real life. Based on the considerations mentioned in the analysis phase, the necessary structures were given collision properties. As mentioned before, the structures that will need this type of property are those whose function is to stop or be an obstacle for the user, taking into account that the vehicle needs them in a mandatory way, since it is the object of training. In addition to this, they had to have considerations like those of gravity, since several aspects will be tested along the levels of the

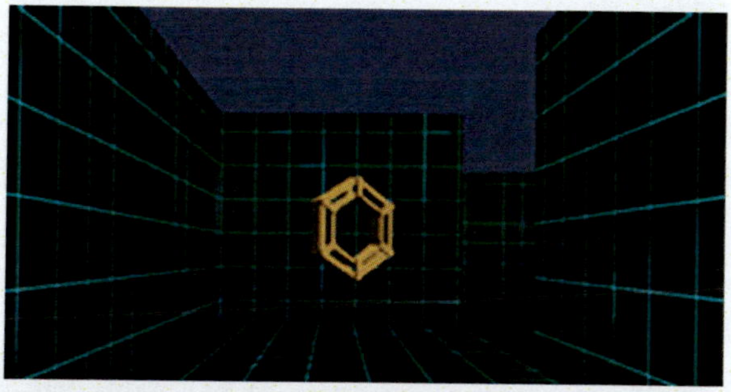

Fig. 12 Texture of the medal compared to the background

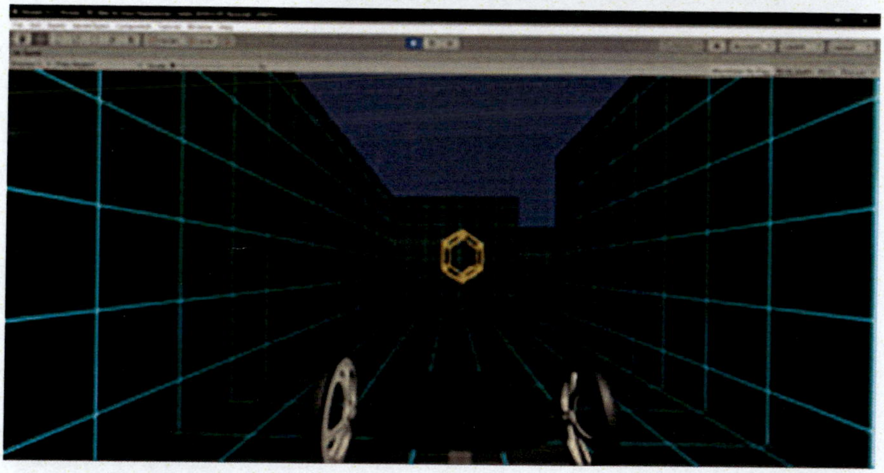

Fig. 13 Illumination and camera settings

simulator, for example, in the levels that have the difficulty that have more complex structures with different levels in the structure or in those that have edges in which the user could fall. Finally, about physics, we worked with the placement of the medals that will be found along the maps. To these, we added an animation that makes them go around, making it easier for the user to identify them.

After finishing with the physics of the simulator, we continue with editing the cameras and lighting. Given the atmosphere of the simulator, we opted for a low lighting to achieve a more striking immersion for the user. As for the work in the cameras, this was anchored so that the user's view was that of a third person perspective, since for a user who has never had a similar experience, the first person could have side effects on the user. Both illumination and camera settings are illustrated in the image below (Fig. 13).

3 Testing

A software product should only be released after it has gone through a proper process of development, testing, and bug fixing. Testing looks at areas such as performance, stability, and error handling by setting up test scenarios under controlled conditions and assessing the results. This is why exactly any software has to be tested. It is important to note that software is mainly tested to see that it meets the customers' needs and that it conforms to the standards [14].

As for the menus, in the simulator, each of the buttons fulfills the expected function, and each of the directions they should have were checked, for example, the level selection menu is opened as soon as the "PLAY" button is clicked.

Regarding the levels or maps, as far as the levels or maps are concerned, we first tested the work of the textures and that they have been placed correctly on each of the objects without causing interference.

After checking the textures in all the structures, the next step was to test how the simulator works in different facets. The first one was to check that the movements of the vehicle were as indicated, for example, that when a joystick or button that is programmed that the movement of the vehicle was to the right accomplished its function. For this purpose, a level was randomly selected and then analyzed the movements by pressing the buttons independently, as well as all the combinations, resulting in all cases was fulfilled the objective because in each of the buttons that were assigned to each direction achieved its task as well as the combinations, for example, forward and right and to change their direction.

The next part of the testing of this project was to test that the physics developed with the scripts were the most appropriate and in the same way that they worked correctly. The goal was to analyze that the given physics work together with the vehicle movements in the simulator to achieve the most realistic experience possible and to achieve a more complete training consequently. Different levels had to be studied to test the falls and the behavior of all the elements that make up a level; for example, if the vehicle falls from an edge to another surface, it would be necessary to check that this surface had the correct properties and that the fall of the vehicle was as close as it would be in real life. Given that several of the maps contemplated to test the user in situations where he must overcome obstacles such as bridges, false paths, etc. have two or more levels in their structure, it was necessary to check how he would behave in the case where there was a surface. It was also necessary to prove that when there was no surface, the vehicle would return to the part from which it fell, and an attempt would be discounted from those the user counted.

Another aspect to consider in the physics was the programmed movements that were considered when programming and developing the scripts. In this analysis, it was found that the medals performed the programmed movement correctly and that the wheels start moving when the vehicle moves forward, so this part also fulfills its purpose.

4 Conclusions

Once the steps that were considered for this project were completed, the corresponding tests were performed, and the results were analyzed, it was concluded that the main objective was partially fulfilled; this is because the operation of the simulator was desired in each of its facets, which makes it a viable option for training, since it tests several aspects to be taken into account in the development of a training software in a correct way. However, the tests conducted do not reflect the expected result, because the main idea is to train a user who would pilot a real prototype and, from this, analyze whether the training that is carried out with virtual reality technology is effective in this area, comparing the real training with this simulator; this will be the scope for the next paper. At this work, it is concluded that the simulator meets all the aspects that were taken into account for the development of it. Although there are details that could be improved to increase the quality of the user experience and the tests have been minimal due to external circumstances, the simulator meets the requirements, and it is estimated that it would be a viable option if at some point a software of this type is required, given that the tests are sufficiently demanding to improve the handling and that it would also require the most experienced pilots.

References

1. M.E. Porter, J.E. Heppelmann, Why every organization needs an augmented reality strategy. Harv. Bus. Rev. **95**(6), 46–57 (2017)
2. S. Benford, C. Greenhalgh, G. Reynard, C. Brown, B. Koleva, Understanding and constructing shared spaces with mixed-reality boundaries. ACM Trans. Comput. Human Interact. **5**(3), 185–223 (1998)
3. M. Kundalakesi, T. Swathi, B. Ashapriya, R. Sruthi, A Study of Virtual Reality. Department of BCA and MSc SS, Sri Krishna College of Arts and Science, Kuniamuthur, Coimbatore, India (2017)
4. R. Silva, J.C. Oliveira, G.A. Giraldi. *Introduction to Augmented Reality National Laboratory for Scientific Computation*
5. E. Gandolfi, *Virtual Reality and Augmented Reality* (2018)
6. A. Tocu, A. Gellert I-R. & Stefan, T-M. Nitescu, G-A. Luca, The impact of virtual reality simulators in manufacturing industry. https://doi.org/10.21125/edulearn.2020.0905 (2020)
7. Abi Research. Virtual Reality Devices to Ship 43 Million Units by 2020, with Mobile-Reliant Head Mounted Displays Foremost, Oyster Bay, New York (2015)
8. M. Suomalainen A. Nilles S. Lavalle. Virtual Reality for Robots (2020)
9. U. Marjanovic, S. Tegeltija, N. Medic, M. Lazarevic, N. Tasic, B. Lalic, Content Development for Virtual Reality Training University of Novi Sad, Faculty of Technical Sciences, Trg Dositeja Obradovica 6, Novi Sad, Serbia (2018)
10. J. Polcar, M. Gregor, P. Horejsi, P. Kopeček, Methodology for Designing Virtual Reality Applications. https://doi.org/10.2507/26th.daaam.proceedings.107 (2016)
11. T. Takala, A Toolkit for Virtual Reality Software Development – Investigating Challenges, Developers, and Users (2017)

12. S. Oveisi, M. Farsi, A. Moeini, Software safety design in requirement analysis phase for a control system (2019)
13. S. Harms, J. Hastings, Enabling Student Innovation through Virtual Reality Development (2016)
14. D. Raghuvanshi, Introduction to Software Testing. 797–800 (2020)

Data-Driven Cyber Threat Intelligence: A Survey of Mexican Territory

M. A. Arturo E. Torres, Francisco Torres Guerrero, and Arturo Torres Budgud

Abstract Information technologies, as well as digital information and information assets, play a very important role today globally, which is why we have witnessed how publications related to cybersecurity incidents are increasing day by day; therefore, and in the face of the great growth of cyber threats, various researchers have dedicated a large part of their efforts to protect these information assets using intelligence sources to develop various techniques for understanding, evolving, detecting, and proactively responding against the cyber threats they face. For their part, companies, governments, and cybersecurity specialists have shown great interest in consuming these sources of intelligence called cyber threat intelligence (CTI), which consists of evidence-based knowledge, which includes context, mechanisms, actionable indicators, implications, and advice on an existing or emerging threat or danger to assets that can be used to inform decisions regarding the subject's response to that threat or danger. With a focus on the Mexican territory, this work aims to analyze the data obtained from CTI sources using the detection of perimeter cybersecurity devices (firewalls, intrusion prevention system, antivirus, honeypots, etc.), as well as the study of research related to cybersecurity predictions to point out the importance of having a model capable of making a possible prediction of cyber threats in Mexico. Challenges and future directions in this field are also discussed.

Keywords Information technology · Cybersecurity incidents · Cyber threats · Intelligence sources · Cybersecurity · Cyber threat intelligence (CTI) · Firewalls · Intrusion prevention system · Antivirus · Honeypots

M. A. A. E. Torres (✉) · F. T. Guerrero · A. T. Budgud
Universidad Autonoma de Nuevo Leon, San Nicolás de los Garza, N.L., Mexico
e-mail: arturo.torrescv@uanl.edu.mx; francisco.torresgrr@uanl.edu.mx; arturo.torresbg@uanl.edu.mx

F. Torres-Guerrero et al. (eds.), *2nd EAI International Conference on Smart Technology*, EAI/Springer Innovations in Communication and Computing,
https://doi.org/10.1007/978-3-031-07670-1_7

1 Introduction

With the constant use and dependence on technology, thanks to innovation and digital transformation, such as the Internet of Things (IoT) or cloud computing, it has become very important to have protection measures for our assets of digital information against cyber threats and/or cybercrime, which Cybersecurity Venture predicts that cybercrime will cost around $6 billion annually by 2021, making it more profitable than global trade in all major combined illegal drugs [1]. Likewise, in the Annual Risk Report 2020 [1] published by the "World Economic Forum" classifies cyberattacks as a latent risk to be prepared from, since the probability and impact on the economy caused by this phenomenon are only below risks such as natural disasters, crises due to lack of water, extreme climates, followed by risks such as infectious diseases, human-made environmental disasters, food crises, etc. In the same study [2], they talk about "the dangers of digital evolution" and how the IoT is also amplifying the potential of the cyberattack surface, estimating that today there are more than 21 billion smart devices or Internet of Things (IoT) worldwide and expected to double by 2025 [2], which have become tools used by cybercriminals, a case that occurred in late 2016, in which they launched a major attack known as distributed denial-of-service (DDoS), causing an interruption in Internet services that affected many companies, including Amazon, PayPal, Netflix, Spotify, and Twitter [3]. Likewise, *Forbes* magazine published [4] that researchers from the cybersecurity company F-Secure detected an increase of more than 300% in attacks on IoT devices in the first half of 2019 [5], while in September 2019, these devices were used to bring down page services such as Wikipedia through a distributed denial-of-service (DDoS) attack [6], and it is estimated that there will be an increase in the use of IoT devices as intermediaries between attackers and their victims.

According to the United States National Institute of Standards and Technology (NIST), cyber risk is defined as the risk of financial loss, operational interruption, or damage, due to the failure of the digital technologies used for informative and/or operational functions introduced to a system by electronic means without authorized access, for the use, disclosure, interruption, modification, or destruction of the systems [7]. The term cybernetic risk, or cyber risk, is closely linked to the concepts of cyber threat and cyberattack.

Given these incidents and the large number of cyber threats hovering around the Internet affecting different sectors of the industry, research has been conducted in sectors such as the health sector [8] which the authors indicate the health sector as a main target of cyberattackers for the theft of personal, critical, and confidential information; likewise, there are also cybersecurity investigations in other sectors such as manufacturing [9] where an investigation of cybersecurity in digital manufacturing systems is presented with a particular focus on the characterization of the system, identification of threats and vulnerabilities, attack scenarios, control methods, and risk determination techniques; we can also find investigations based on financial market reactions to a cybersecurity attack [11].

In these situations, the various communities, as well as manufacturers and cybersecurity researchers, have shown great interest in publishing their findings and research to the general public, resulting in a large amount of information that can be consumed by analysts to make important decisions about when carrying out a cybersecurity strategy, such as allocating resources, budgets, and prioritizing actions in the face of a cyber threat [10–12]. However, as of today, there are few threat prediction works that focus on perimeter security schemes based on the data collected by these devices.

The most widely used techniques to perform cybersecurity prediction focus on obtaining cyber threat intelligence (CTI) sources of information which consists of obtaining a data set of cybersecurity events to process and analyze with techniques such as data mining (DM) and machine learning (ML) to be able to make a decision based on the events or data obtained previously. Therefore, in a scenario where cyber threats can be predicted, any company and/or information technology providers, as well as users, could inform themselves and protect themselves from the impacts caused by cybersecurity threats.

And as we know, no system or device is perfect or can be considered 100% secure, and in the face of the great growth of everyday devices that connect to the network plus the dependency that we are generating towards them, we can consider that cyber threats are a constant that we must take into account at any time that we use a service or technology. That is why various studies have been carried out on the advances and developments of prediction of cyber threats and incidents, highlighting the study of [15], in which the authors present the compilation and investigation of schemes, methods, and data sets for predicting cybersecurity incidents of the latest generation, highlighting the existing work in this field. Likewise, the authors organize their research into six categories, according to the data sets used, such as reports and data sets of the organization, network data sets, synthetic data sets, web page data, social network data, and mixed-type data. Therefore, proactive prediction of cybersecurity threats and incidents is considered a potential and immediate problem to be solved. That is, the prediction of cyber threats is an area of research where there is a large area of opportunity to be studied and developed.

1.1 Contributions

In this research article, it is organized as follows:

- Section II aims to present the investigation of the current state of cybersecurity in Mexico in order to obtain an overview.
- The main contribution of Section III is the compilation and investigation of CTI schemes, methods, and data sets corresponding to the Mexican territory available to analyze in future investigations.
- Section V discusses the challenges and opportunities for future research in this area.
- Section VI presents the conclusions of the research article (Fig. 1).

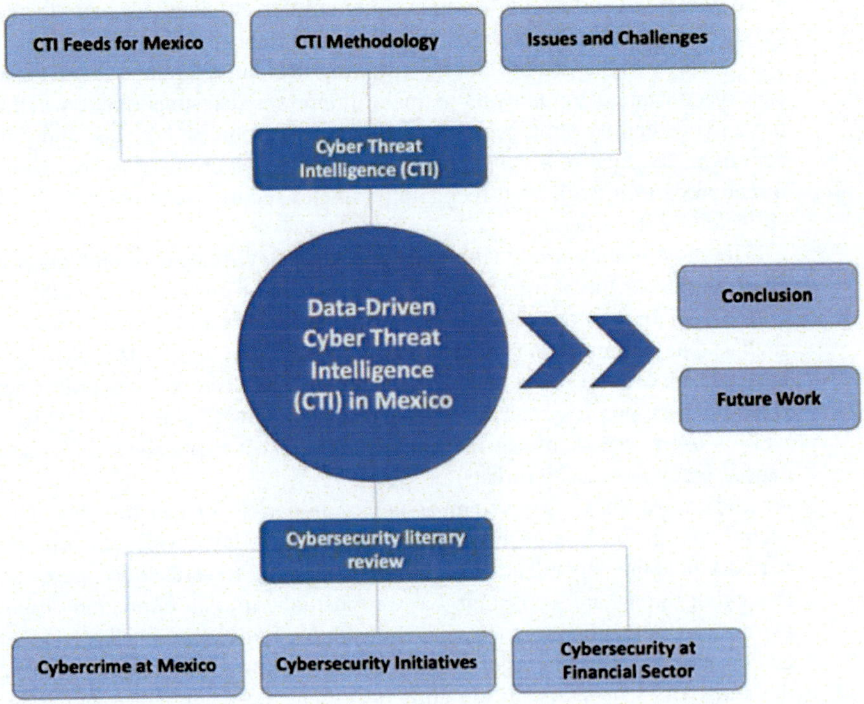

Fig. 1 Methodology

2 Current State of Cybersecurity in Mexico

In this section, a general vision of the current cybersecurity situation in the Mexican territory is presented, which aims to cover the different articles issued by consultancies in conjunction with the Mexican government, as well as studies on different sectors of the industry to have a broad overview of current cybersecurity in the country. For this study, we will use the term cybersecurity defined by the ISO/IEC 27032 standard as the preservation of the confidentiality, integrity, and availability of information in cyberspace, which in turn is defined as the complex environment that results from the interaction of people, software, and services on the Internet through technological devices and networks connected to it, which do not exist in any physical form [13].

Mexico is a country in constant industrial and technological development; however, there are certain emerging issues in the new industrial policies, which go beyond the traditional areas of promotion, protection, and industrial and service regulation. The Scientific and Technological Advisory Forum [14] published a document where issues such as the efficient use of energy and sustainable development are pointed out; competition in the national and international market; education and training; and promoting scientific research and technological development [15]. In this same document, it is mentioned that said developments have been characterized by the dominance of foreign capital and technology and the low national added value. In contrast to what happened in China, South Korea, India, and other Asian countries, companies with national capital have not co-invested in Mexico with foreign companies, nor have they excelled in the development of their own technologies, or their exports (with the exception of the mining sector). Likewise, the great challenge of generating a successful economy where it is only based on foreign investment is mentioned, as has been understood in Mexico and other Latin American countries in recent years.

Cybersecurity in Mexico is an issue that has grown strong in recent years, and the cyberattacks that we have faced recently have had a significant national impact, and the media are increasingly taking more value on news related to the cybercrime [16, 20, 21]. In summary, it affects in some way all sectors of the country due to the rapid growth made up of technology, social networks, information systems, and the Internet, where Mexico achieves a 71% penetration among the population of people 6 years and older. With 79.1 million users connected, therefore, Mexican companies have adapted new business models, such as electronic commerce, in which it was revealed that 8 out of 10 Internet users of legal age have made a purchase online in the last year, in which 85% of online purchases in Mexico are made through smartphones, highlighting that already 2 out of 10 online buyers make some purchases on their Smart TV, reaching a value of more than 491 billion pesos and with a growth of 24% compared to 2017 [17]. This tells us about the adoption and constant use of digital media to consume a service or purchase a product in the Mexican territory where Internet users in Mexico spend 8 hours 20 minutes daily, 8 minutes more than in 2018 [23].

Almost all aspects of daily life in the country depend today on the use of information and telecommunications technologies, which favor and improve the lives of Mexicans as well as improve productivity by being able to improve and automate processes. As evidence of this, there has been an exponential growth in the use of devices connected to the Internet, and it is estimated that by 2025 there will be more than 300 million devices with access to networks in Mexico; this is 70% more than the 180 million documented in 2018 [18]. The increasing dependence on information technologies and the accelerated increase in cyber threats as well as cybercrime have forced Mexican companies in the public and private sector to increase their investments and budgets in strategies and cybersecurity controls with the main objective of being able to protect from these threats that affect the availability, confidentiality, and integrity of your information and your business. The financial sector is a critical system that has a high dependence on information technologies, since its transactions between clients, other institutions such as businesses and electronic payments are carried out digitally and in which studies have been carried out where the impact that occurs in a company when it is the victim of a cyber threat [11] it is shown. Likewise, different sectors of the industry have also adopted technology as a day-to-day tool to automate and optimize many of their critical processes where we have mentioned various scientific articles, which seek to improve their detections and incident response process according to their sector [8, 9].

In February 2019, the Secretary of Communications and Transportation (SCT) [19] in conjunction with the Organization of American States (OAS) [20] published a study on the habits of users in cybersecurity in Mexico, in which more than 5 thousand people residing in different states of the republic were interviewed and in which the annual increase in users on the Internet stands out exponentially with a growth of 4.7% from 2015 to 2016 and an increase of 8.1% from 2016 to 2017. Where free Internet access is identified, as well as mobile device applications for minors as one of the most important concerns, since according to the study only 45% of adults monitor the content minors visit and/or consume and 37% declare the use of these mobile devices and applications as entertainment for their children without controlling the content or pages they relate to while accessing the Internet. Given this annual rise, immediate action is paramount regarding day-to-day activities since the study revealed that 42% of the participants indicated not knowing the permits required by the applications before installing them as well as more than 20% of the users accepted being victims of financial fraud through digital means mostly by email [21] using techniques such as phishing, which is defined by the Federal Police as a type of scam, whose objective is to obtain data, passwords, bank account numbers and credit cards, identity, or other data to be used fraudulently [22].

In the document issued by the McKinsey & Company consultancy in collaboration with COMEXI, cyber risks are pointed out as a new threat that we must face. Likewise, they define the concept of "cyber risks" as a set of possible damages that companies, governments, and members of society could suffer due to a failure or violation of the information technologies they use every day. This can be reflected in an impact of economic loss, damage to the reputation of a company or person, as well as in the loss of availability to provide a service or in making poorly informed decisions [18]. These can occur in various ways, such as a system error or vulnerability, some accidentally caused failure, or configuration error; however, the greatest damages usually arise from a cyberattack directed either by an external actor or some internal actor of the company. The term cyberattack is defined as an unauthorized attempt by some digital way to access a system, information, and/or resource in order to exfiltrate the information, compromise it, and affect its availability until it ends up extorting users and organizations and corresponds to the materialization of one or more cyber threats.

2.1 *Cybercrime in Mexico*

As mentioned throughout this document, in recent years, the evolution and technological dependence at a global level has had an accelerated growth, which although it offers us many advantages also carries risks for the daily use of technology and information that we enter to be able to use it, for example, an application to go to the movies, buy a pantry or food or transport, you can ask for personal and financial information to be able to use it and even administrator permissions on your devices, access to your location, photos, and others; another example is the exponential use of social networks, which today is estimated that an average user in Mexico spends up to 8 hours connected to the Internet and uses more than three social networks (Facebook, Twitter, Instagram, Twitch, TikTok, etc.) in a day [23]. This added to the lack of security measures by users when entering unknown sites or installing applications without reading or understanding the permissions or risks involved, increases the risk of contracting an infection and spreading some type of computer threat within of its devices that can trigger a cybercrime having as a victim some user or institution.

The term cybercrime or cybernetic crime is a form of crime that uses both the Internet and technology as a means of committing some unlawful act that may harm integrity, confidentiality, or availability. Some problems related to this type of crime are fraud, information hijacking (ransomware), phishing, malware distribution, distributed denial-of-service (DDoS) attacks, piracy, child exploitation, information theft and/or identity theft, as well as privacy breaches when confidential information is lost or stolen; they become increasingly common in our society, and this has alerted governments and organizations globally to increase their investments in cybersecurity controls to deal with it. According to the generally accepted international classification, we present some of the terms and concepts of cyber threats defined by the NIST [23].

Cyber threats	Description
Malware	Simplified term to denote "malicious code" and consists of software intended to carry out an unauthorized process that will have an adverse impact on the confidentiality, integrity, or availability of an information system. Within this category are mainly the following types
Virus	Hidden and self-replicating section of computer software, which spreads by infecting (i.e., by inserting a copy of itself into another program and becoming part of it). A virus cannot run alone; requires your host program to run to activate it
Spyware	Software that is secretly or surreptitiously installed in an information system to collect information about individuals or organizations without your knowledge
Adware	Software that automatically plays, displays, or downloads advertising material to a computer after installing the software or while using the application. The malicious program is designed to display unwanted advertisements on the victim's computer without their permission, pop-ups or advertisements are uncontrollable and tend to behave erratically, they usually appear many times on the screen, and it is tedious to close them

(continued)

Cyber threats	Description
Rootkit	A set of tools used by an attacker after gaining root-level access on a host to hide the attacker's activities on the host and allow him to maintain root-level access to the host through secret means. In other words, it allows a hacker to remotely access or control a computing device or network without being exposed. They are difficult to detect because they are activated even before the operating system starts
Trojan horse	Computer program that seems to have a useful function, but also has a hidden and potentially malicious function that evades security mechanisms, sometimes exploiting legitimate authorizations from an entity that invokes the program
Worm	It is the simplified term to denote "write once, read many"; it consists of a computer program that can be run independently, can propagate a full version of itself to other hosts or networks, and can destroy a computer's resources. In other words, it is malicious code that is also copied and spread to other computers, a system, or a network
Ransomware	It is a virus that prevents the user from accessing the files or programs, and for its elimination, it is required to pay a "ransom" through certain online payment methods. Once the amount is paid, the user can resume using their system
Keylogger	A program designed to record which keys are pressed on a used computer keyboard, to obtain passwords or encryption keys
Botnet	It is a network of devices that has been infected with malicious software, such as a virus. Attackers can control a botnet as a group without the owner's knowledge in order to increase the magnitude of their attacks. Often a botnet is used to overwhelm systems in a distributed denial-of-service (DDoS) attack
Phishing	A technique of trying to acquire sensitive data, such as bank account numbers, through a fraudulent request in an email, or on a website, in which the perpetrator is posing as a legitimate business or reputable person
Man-in-the-middle attack (MitM)	A MitM attack is when an attacker disrupts communication between two users, posing as both victims to manipulate them and gain access to their data. Users are not aware that they are actually communicating with an attacker and not with each other
Distributed denial-of-service (DDoS) attack	A denial-of-service attack floods systems, servers, or networks with traffic to drain resources and bandwidth. As a result, the system cannot fulfill legitimate requests. Attackers can also use multiple compromised devices to launch this attack. This is known as a distributed denial-of-service attack
SQL injection	It occurs when an attacker inserts malicious code into a server that uses Structured Query Language (SQL). They only succeed when a security vulnerability exists in the software of an application. Successful SQL attacks force a server to provide access or modify data
Zero-day attack	An attack that exploits a previously unknown hardware or software vulnerability. Using outdated (unpatched) software opens up opportunities for criminal hackers to exploit vulnerabilities. A zero-day vulnerability can occur when a vulnerability is made public before the developer has implemented a patch or solution

Cybercrime has become one of the most important security issues that will continue to emerge as a critical problem for years to come. Among the different attacks, the use of techniques to exploit a vulnerability is of special interest due to its negative impact on the economy. In the last decade, cybercrime has transformed from a low-volume crime to a high-volume crime. During that time, the perpetrators have switched from specialized individuals to expert attackers who have established organized structures to carry out some structured cyberattack. Likewise, cyberattacks are considered one of the risks with the greatest impact in the coming years, according to the "Global Risk Report 2019," which presents the results of the last "global risk perception survey," in which nearly a thousand decision-makers from the public sector, private sector, and academic and civilian society evaluate risks faced by the world [24].

In Mexico, various studies related to cybercrime have been carried out, explaining that this phenomenon has become something much more complex and of greater impact, where it is estimated that more than 80% of Mexican companies are victims of cyberattacks by at least once a year, which positions Mexico as one of the ten countries with the highest number of cyberattack attempts and cyber threats globally [25]. In 2017, 33 million Mexicans (50% more than in 2016) were victims of some cybercrime-related cyber threat, that is, one in four inhabitants of the country. At the same time, the economic impact of these crimes is estimated to have amounted to 7.7 billion dollars, 40% more than the previous year [18], being human behavior (negligence or malicious acts of workers) one of the main factors of these cyber risks.

In Mexico, we have witnessed an increase in cyber threats that have affected different sectors of the industry in recent years, such as the one that affected the National Electoral Institute (INE) in 2016, where it was possible to violate a hosted database in the Amazon Web Services (AWS) cloud, in which data from more than 93.4 million Mexicans was exposed [26]. Another case in 2018 where it was revealed that some cyberattackers violated some financial institution systems that interacted with the Electronic Payments System (SPEI), which resulted in the theft and loss of approximately 300 million pesos from different locations, in which the attackers created fake accounts to send fraudulent payment instructions through a malicious code, tricking financial institutions to send transactions from Banco de Mexico through SPEI, as it usually does. As a result of this, various companies were affected, such as AXA Seguros, which reported inconsistencies related to the payment system, and its operative field declared that the attack was aimed to their connection systems with the SPEI platform causing an economical loss of approximately 57 million pesos [27, 34].

As detailed in the McKinsey & Company document in collaboration with COMEXI [18], there are groups called hacktivists for the purposes of carrying out a cyberattack with the main objective of making a political declaration or protest towards a government or institution. In 2012, the Cyber Protesta Mexicana group carried out a simultaneous cyberattack, where protest messages were published on at least 10 government, political party, and press websites. This group, which had similarly attacked in 2009, positioned it cyberattack as a peaceful protest against the country's political situation, in which its activities have become global notes [28]. In Mexico alone, during the fourth quarter of 2019, cyber fraud increased 36% compared to the same period of the previous year, representing an approximate amount of 11,171 million pesos, and only 45% were rewarded and 86 out of 100 cyber frauds were able be resolved in favor of the affected user according to the document prepared by the Comision Nacional para la Proteccion y Defensa de los Usuarios de Servicios Financieros (CONDUSEF) based on the claims with monetary impact presented by the clients of Banco de Mexico contained in the Regulatory Report R27 (CNBV) [29].

According to several studies carried out since 2019 by Fortinet and its research laboratories, FortiGuard, Mexico is the country that presents the highest number of cyberattacks in Latin America [30]. This represents a great challenge for Mexican users and companies that face this new wave of digital crime every day.

2.2 *Cybersecurity Initiatives in Mexico*

Although different agencies can independently implement cybersecurity initiatives, the organization and integration of these efforts requires an agency that can properly manage them. Given this, in 2017, the government of Mexico issued the National Cybersecurity Strategy [31], which establishes the vision of the Mexican State in the matter, taking into account the impotence of information and communication technologies (ICT), the risks associated with the use of technologies and cybercrimes, as well as the need for a general culture of cybersecurity. The general objective of this strategy is to identify and establish cybersecurity actions applied to the social, economic, and political spheres that allow the population and public and private organizations to use and take advantage of ICT in a responsible manner for the sustainable development of the Mexican State.

On the other hand, the Mexican legal framework also typifies cybercrime, although in a decentralized manner, however, there is a legal framework for the protection of personal data by the Federal Law on Protection of Personal Data Held by Private Parties (LFPDPPP) [32], which specifies in its Article 7 that the collection of personal data should not be done through deceptive or fraudulent means, as well as Article 20, which dictates that security breaches occurred at any stage of the treatment that affect in a way the proprietary or moral rights of the owners will be informed immediately by the person responsible to the owner, so that the latter can take the corresponding measures to defend their rights and Article 58. The owners who consider that they have suffered damage or injury to their property or rights as a result of noncompliance with the provisions of this law by the person in charge or the responsible may exercise the rights that they deem pertinent for the purposes of the corresponding compensation in terms of the corresponding legal provisions. There is also a General Law on the Protection of Personal Data in Possession of Obligatory Subjects (LGPDPPSO) [33] which contains specific topics on technological issues, cloud computing, and security specified in Article 3, where technical security measures are defined as a set of actions and mechanisms that use technology related to hardware and software to protect the digital environment of personal data and the resources involved in its treatment. As well as in the LFPDPPP it is also specified in Article 19. The person responsible must not obtain and process personal data, through deceptive or fraudulent means, privileging the protection of the interests of the owner and the reasonable expectation of privacy. These legal frameworks have as their main objective the obligation of every entity, which handles personal data, to establish and maintain technical and physical administrative security measures. These allow to protect personal information against damage, loss, destruction, use, access, or unauthorized treatment. The legal framework also specifies confidentiality obligations, which define the specific cases in which private information may be shared with other entities, and the precautions due for that transaction. For the implementation of data security measures, those responsible must consider the existing risk, the possible consequences for the data owners, and the technological development that is available. The National Institute for Transparency, Access to Information and Protection of Personal Data (INAI) develops and publishes material to assist organizations that handle personal data and thus fulfill its responsibilities. Any organization that does not adhere to this legal framework is subject to sanctions, but these can be reduced if the authority considers that the organization followed the recommendations of the INAI.

It should be noted that a cyberattack is not considered a crime in the legislation of our country, that is, only those that are considered computer crimes, since there is no legislation on cybersecurity, for example, the theft of wireless networks (Wi-Fi) is not considered a computer crime; however, there are some that can be reported, for example, theft of data and personal information or cyberbullying [34, 35]. Given this, the Government of the State of Mexico, through the Secretariat of Security of the State of Mexico, created the Cybernetic Prevention and Investigation Unit or Cyber Police, whose main objective is to prevent, attend, and combat incidents that are committed through digital means, such as fraud, extortion, identity theft, sexual exploitation, harassment, animal abuse, and sale of prohibited substances and weapons, among others. This unit works 24 hours a day, 365 days a year, through its three areas of operation: citizen attention and cyber patrol, technological laboratory, and prevention of cybercrime [36]. On the other hand, in Mexico, some of the main universities already offer careers or educational programs in cybersecurity (e.g., UNAM, ITAM, ITESM, IPN, UNITEC). It should be noted that the Autonomous University of Nuevo León (UANL) offers the Bachelor of Security in Information Technology in which agreements have been made with the cybersecurity firm Fortinet to include its Fortinet Network Security Academy (FNSA) program, which provides academic institutions (high schools, colleges and universities, and nongovernmental organization (NGO) focused on career readiness), with the resources necessary to facilitate Fortinet's industry-recognized certification curriculum.

2.3 Cybersecurity in the Mexican Financial System

In Mexico, cybersecurity strategies have only recently been agreed for the financial sector and for the country in general, due to the fact that in 2017, Mexico ranked seventh among the 20 most important markets in the world in terms of adoption of FinTech companies, defined as the proportion of the adult population that has used at least two services of companies classified as FinTech in the last 6 months [37]; this makes for a more efficient and inclusive sector than that of some years ago; however, in addition to the growth of new industries, such as FinTech, the risk of cyberattacks on the financial sector has also increased. Given the accelerated increase in cyberattacks on the financial sector, in October 2017, the Cybersecurity Forum "Strengthening Cybersecurity for the Stability of the Mexican Financial System" was held [45]. This forum brought together authorities, representatives of financial institutions, and national and international experts to analyze best practices at the international level to strengthen cybersecurity measures, set an agenda, and coordinate efforts between authorities and the private sector, and communicates the Principles for Strengthening the Cybersecurity for the Stability of the Mexican Financial System [38]:

1. Adopt and keep updated policies, methods, and controls to identify, evaluate, prevent, and mitigate cybersecurity risks, which are authorized by the highest decision-making governing bodies and permeate all levels of the organization.
2. Establish secure mechanisms for the exchange of information between the members of the financial system and the authorities, about attacks occurred in real time and their mode of operation, response strategies, new threats, as well as the results of investigations and studies, that allow entities to anticipate actions to mitigate the risks of cyberattacks; the foregoing, protecting the confidentiality of the information.
3. Promote initiatives to update regulatory and legal frameworks that support and converge the actions and efforts of the parties, considering best practices and international agreements.
4. Collaborate in projects to strengthen the security controls of the different components of the infrastructures and operating platforms that support the country's financial services, promoting the use of information technologies to prevent, identify, react, communicate, typify, and make a common front in the face of present and future threats.
5. Promote cybersecurity education and culture among end users and the staff of the institutions themselves that, through continuous training, result in active participation to mitigate the current risks of cyberattacks.

However, the National Banking and Securities Commission (CNBV) in collaboration with the Organization of American States (OAS) [39] revealed that 100% of Mexican financial entities and institutions affirm that they identified some type of digital security event, that is, successful attacks and/or failed attacks that attempted to violate them. The most commonly identified digital security events during 2018 were malware with 56%, targeted phishing with 47%, and breach of security policies with 31%. Likewise, it is highlighted that 19% of financial entities and institutions daily identify the occurrence of events of some type of cyber threat, taking into account one of the main reasons why attacks are carried out in this specific sector. They are mainly for economic reasons. Likewise, in this study, it is mentioned that the main cybersecurity actions carried out by Mexican financial institutions consist of implementation of controls such as firewalls (85%), automated antimalware consoles (76%), automated backups (68%), and network security (VPN, NAC, ISE, IDS/IPS, web filtering, secure email, etc.) (54%). Therefore, we can assume that there is a great opportunity for development in the area of research and development in technological and cybersecurity fields in Mexico.

3 Cyber Threat Intelligence

Given the accelerated growth of cyberattacks, as well as the constant development and use of new digital platforms, organizations face more complex challenges every day when it comes to monitoring their information, that is, each application, device, or process that interacts with technology generates digital information that each, in turn, represents an event (log) in a system that cybersecurity analysts should take into account to monitor any type of suspicious activity; however, as we have mentioned, the constant use, evolution, and dependence of technology have resulted in an impressive amount of information, including possible threats that in many cases exceed analysts' analytical skills and risk omitting or reporting this information as noise. Likewise, cyber threats are becoming increasingly complex, developing new ways to violate systems using new techniques, as well as the ease of use of tools that allow any user to be able to generate these types of threats with very little effort and cause a high impact. Given this, the main challenge lies in being able to transform all this information into something actionable that can be used to make decisions for senior management, for example, being able to prioritize activities and allocate budget or personnel based on the impacts that can be had based on the collected data. This requires not only the time and/or effort of the IT or cybersecurity team but also organization, collaboration between the different areas, experience, and resources allocated by senior management to carry out these investigations successfully. This leads us to what is called cyber threat intelligence (CTI), which is defined as evidence-based knowledge, which includes context, mechanisms, indicators, implications, and practical advice, about an existing or emerging threat or danger for assets that can be used to inform decisions regarding the subject's response to that threat or danger [40]; on the other hand, SANS [41] defines CTI as analyzed information about the capabilities, opportunities, and intentions of adversaries that meets specific requirements determined by an interested party.

3.1 CTI Processes and Methodology

One of the main points that needs to be understood in this area is the difference between data, information, and intelligence to understand CTI. Therefore, we can define data as an individual element that contains information, either from a system, action, or executed process, that is, individual elements with a specific meaning. On the other hand, we can define the term threat as the possible danger that can be used to exploit an existing vulnerability with the intention of causing damage to systems, networks, or entire organizations. CTI is defined as information on how to detect and defend against cyber threats by aggregating data analysis and evaluation information with meaning.

The CTI cycle is a process to generate accurate and actionable information for the organization [41]. It begins with a planning phase, in which the intelligence questions or requirements that need to be answered are generated. When the requirements are known, the next phase is collection, the collection of data to help answer the questions and meet the requirements. The next phase is processing, where the data is put into a usable format for analysis. This leads to the fourth phase, analysis, in which the data is synthesized to identify responses to intelligence requirements. The last phase is dissemination, where the findings are captured in the correct format to reach the intended audience described in the planning phase. It is important to keep in mind that although the intelligence cycle is a cyclical process, sometimes it is necessary to go back in the process; for example, if during the analysis phase it is determined that additional information is needed or if the information should be processed in a different format, it is important to return to the appropriate previous step so that the final result is an informed and accurate analytical finding. So, as an example of where and how intelligence really adds value to the organization, it will basically give us visibility. It will allow the threat intelligence analyst to add value very quickly, timely, and actionable information that can be fed to defense teams across various roles within the organization and to IT operations people to execute and configure any necessary controls to reduce the risk of an incident and/or cyber threat. Therefore, organizations with CTI processes focus on understanding the cyber threats they face with the primary goal of providing actionable and valuable information to help defend against those threats that put the organization's information at risk.

3.2 CTI Mexico Feeds

As mentioned above, the key to be able to carry out a good CTI process is the definition of sources and collection of information; that is, once the requirements have been identified, the next step is to identify how to obtain access to the information that will help answer to the requirements. These sources of information are called intelligence feeds, which emit relevant information for analysts to consume it in different ways. In this same article, SANS [41] shows which are the intelligence sources that organizations consume the most in order to carry out CTI.

Sources for gathering intelligence	2020
Open-source or public CTI feeds (DNS, MalwareDomainList.com)	74.30%
Threat feeds from CTI-specific vendors	68.90%
Threat feeds from general security vendors	68.50%
Community or industry groups such as Information Sharing and Analysis Centers (ISACs) and Computer Emergency Readiness Teams (CERTs)	68.20%
Security data gathered from our IDS, firewall, endpoint, and other security systems	63.40%
External sources such as media reports and news	63.10%
Incident response and live forensics	63.10%
SIEM platform	62.00%
Vulnerability data	60.60%
Network traffic analysis (packet and flow data)	57.00%
Forensics (postmortem)	56.40%
CTI service provider	45.90%
Application logs	44.40%
Other formal and informal groups with a shared interest	43.30%
Closed or dark web sources	42.10%
Honeypot data	29.90%
Shared spreadsheets and/or email	21.00%
Other	1.50%

Sources for Gathering Intelligence [41]

This definition can be explained with an example from the perspective of the analysis of cyber incidents in such a way that the data can be such as the IP, the domain, the URL, or the email that can be collected from the systems or sources of information opened on the Internet like Google. In addition, the information can be described as the exploited URL for phishing, the domain that spreads the malicious code, and the IP that establishes C&C communication with the malicious code [50]. Cyber threat intelligence is the result of comprehensive analysis reporting that a group of cybercriminals is targeting primarily financial entities, and it was recently discovered that malicious code was a variant of some previously seen threat. Therefore, actions are required to block the IP address of the C&C server frequently used by malicious code. There are different types of tools that can provide very relevant information to organizations regarding feeds for CTI, which aim to be able to carry out the mentioned intelligence phases, to deliver information already processed to organizations with the main objective of sharing information, cybersecurity studies, and findings to help organizations protect their information and minimize the risk of a vulnerability. Given this, organizations and/or companies dedicated to cybersecurity have adopted the use of CTI in their tool sets and are complemented by products or services that provide information already digested or

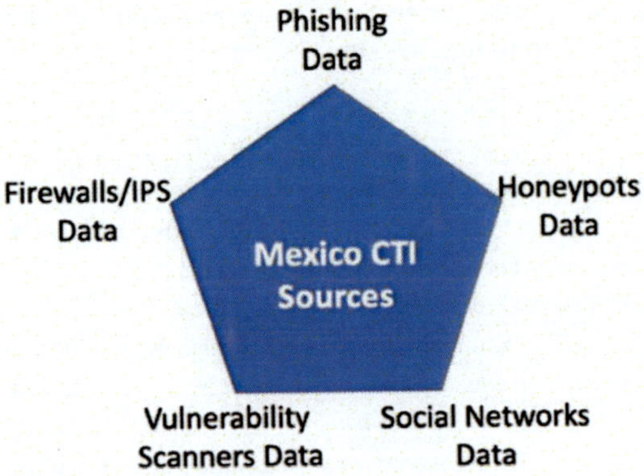

Fig. 2 CTI sources for Mexico

consumable by CTI [42–44], including social media platforms such as Facebook, has launched its ThreatExchange project [45] to share information about CTI, as well as studies related to the extraction of CTI through Twitter [46, 47]. One of the most relevant phases of the intelligence life cycle [41] focuses on being able to collect information related to relevant threats according to the population to be protected in order to understand the risks, threats, and requirements to be investigated; therefore, we will mention some of the platforms that can be used to collect relevant information to Mexico.

As mentioned above, there are various platforms and/or services available to collect relevant information on threat trends, a specific threat, as well as the techniques they use; but what intelligence sources are available that provide information for the Mexican sector? In this document, the contribution is to show some of the platforms that provide information about cyber threats that occur in Mexico and that can be used to generate relevant and actionable information in an organization, that is, using the information collected as a starting point of any malware trend or cyber threat that has occurred or that is targeting Mexico, categorizing them into five sections (phishing, firewall/IPS, honeypots, vulnerability/scanners, and social network data) so that organizations can carry out an investigation on said cyber threat to understand its objective, techniques, and procedures to measure its impact on the business, with the main objective of converting this information into a cybersecurity strategy to reduce the risk of any incident to the organization (Fig. 2).

For this, we will cover some of the main areas of intelligence interest; among them there are cyber threat trend platforms based on detections of cybersecurity events from perimeter devices (firewalls, IPS, EPP, etc.), such as Fortinet Threat Intelligence Insider Latin America [48], a quarterly threat trend tool by FortiGuard Labs [42]. For ten countries in the Latin American region, including Mexico, which has data collected and analyzed from millions of daily cybersecurity events detected by sensors deployed in the region, it also offers data on cyber threat trends by country and information on the ten main cyberattacks for the countries of the region, in the category of malware, exploits and botnets, as well as regional executive summaries of the main areas of risk and vulnerabilities identified, as well as security tips and key findings, in addition to the possibility to download this information in PDF format, which is updated quarterly in the three main languages of the region (English, Portuguese, and Spanish). Another CTI source that offers information on vulnerabilities and visibility of devices exposed to the Internet in the Mexican sector is the Shodan search engine [49]; that is, if a device is directly connected to the Internet (from small desktop computers to nuclear power plants, etc.), Shodan consults it to obtain the public information available on that device, in which various studies about the risks that exist on devices exposed to the Internet have been carried out [50]. Shodan recently added panels and exposition of Internet availability for certain countries, including one for Mexico [51], which shows the number of industrial control systems exposed in the country, as well as the most relevant vulnerabilities. In addition, a general search was made in this tool using the "country:" MX "" filter, which gave us a total result of 4,880,789 devices exposed to the Internet [52], of which cities such as CDMX (616,591), Zapopan (212,236), Guadalajara (189,331), Monterrey (180,109), and San Luis Potosí (88,374) stood out.

On the other hand, there are various platforms made up of a network of honeypots, which are nothing more than devices exposed to the Internet that work as decoys to attract cyber threats and to monitor the techniques used by these cyber threats, such as Bad Packets Cyber Threat Intelligence [53], which has a global network of honeypots that detect the activities of active botnets, which are scanning the Internet and/or participating in malicious activities, also have honeypots deployed in Mexico [54], which can provide information on cyber threats that target our country, in which said tool has been used for different studies to monitor botnet campaign traffic by scanning devices exposed to the Internet [55] and the profiling of critical industrial systems (ICS) exposed on the web [56]. Likewise, platforms were also found that provide intelligence feeds related to fraudulent websites, better known as phishing, such as APWG [57]who mention in their Phishing Activities and Trends report for the first quarter of 2020 that country code domains (ccTLDs), such as .UK for the UK and .MX for Mexico, were approximately 44% of the domains in the world at the beginning of the first quarter, but only 27% of the domains in the first quarter sample. Another platform from which phishing information can be downloaded is OpenPhish [57], which offers different plans for the use of its platform, from downloading URLs for free to detailed information on each site. It should be noted that using the free phishing site download offered by the platform [58], at least 15 sites with the domain ".mx" and 2 with the word "mexico" were found in the URL that have been detected with some malicious activity. Another tool that can provide intelligence on malicious sites for Mexico is URLhaus, which offers feeds of malicious sites that can be downloaded, and for Mexico, 291 URLs were found classified as phishing or malicious [59].

Social media platforms allow users and organizations to communicate and share information. For security professionals, it could be more than just a network tool; that is, it can be an additional source of valuable information on topics from vulnerabilities, exploits, and malware to threat actors and anomalous cyber activities. For example, Twitter and Facebook are not just a platform for sharing content, promotion, or social networks. There are open-source intelligence tools (e.g., TWINT [60] and ThreatExchange [61]) that can scrape data or publicly available Twitter streaming application programming interfaces (APIs) that can collect sample data for analysis. We also saw that bots shared the latest indicators of compromise (IoC) and even threat detection rules. In fact, there is publicly available information on how Twitter bots can be used to monitor Internet of Things (IoT) devices. There are also open-source honeypots that can log data on Twitter.

There are cloud-based threat intelligence platforms, such as the IBM X-Force Exchange [62] that allows you to use, share, and act based on threat intelligence. This platform allows you to quickly investigate the latest cyber threats in the industry based on tags, hashtags, or indicators as well as adding actionable intelligence or collaborate with other cybersecurity analysts; it also shows us different options for malware, analysis, profiles, etc. On this platform, 35 cyber threats were found by performing a search with the tag "Mexico" which allowed us to carry out a search for 35 cyber threats documented on this platform, of which only 12 were related to Mexico in 2020. On the other hand, cybersecurity organizations have created real-time cyber threat maps, to provide an overview of the attacks and their relationship between countries; for example, Live Cyber Threat Map is a free Check Point page that shows malware attacks in real time, phishing and exploit in the world, as well as statistics. By focusing on Mexico (by clicking on the map) [63] shows us the trends of cyber threats in the last 30 days: banking Trojans (1.2%), botnet (6.2%), cryptominer (3.6%), mobile (6.5%), and ransomware (0.3%). Another threat map that can provide relevant information for Mexico is the A10 company DDoS map [64], which indicates that in Mexico there are more than 115,000 devices that can be used as weapons to carry out a distributed denial-of-service (DDoS) attack, 21,097 hosts identified and infected with DDoS malware (called drones), 2,884 identified hosts that are carrying out malicious activity (called abuse), 5,760 publicly exposed and vulnerable DNS servers to be exploited by an amplification attack, as well as NTP servers (1,608), SSDP (26,421), SNMP (27,375), and TFTP (28,527), among others. Additionally, a service was found that its main function is to be a search engine for servers and services as its support for IPv6 generating intelligence; with this information, it obtains an analysis and evaluates the risk exposure of organizations in real time [65, 66].

4 Conclusion

This document compiled the most relevant information, as well as the current situation in Mexico using, for the most part, documents issued by the Mexican government. This attracts attention, since there are various studies, initiatives, and strategies under development in the country. Likewise, it was found that performing a CTI strategy requires great demand and constant development by any organization, whether it involves personnel or technology with advanced analysis and detection capabilities, as well as a framework of references and processes to obtain the knowledge for the prevention of cyber threats with the main objective of obtaining evidence that can positively influence the decision-making of the organization; therefore, experience, skills, and information sources of a security team define its ability to produce accurate and actionable cyber threat information [67]. Given this, we were able to find different intelligence sources related to information for Mexico, which will be analyzed in future work and dictate the ability to generate a threat detection model based on the evidence collected from the sources described in this document. It is also clear that there are numerous positive trends in the community, such as more organizations producing intelligence rather than just consuming it. But there are also many challenges, such as getting the right staff and training to conduct cyber threat intelligence. Tools and data sources will always be vital to the process, but the world of intelligence analysis is intrinsically analyst driven, and an approach is rightly placed there. Sharing not only adversarial IoC and TTP, but also analytical processes, will help the community continue to grow. Some sharing processes include strategies to measure the effectiveness of a CTI program [41].

4.1 Challenges and Future Work

The next steps in this and future investigations include identifying the intelligence sources necessary to carry out a predictive model based on data and evidence collected for the Mexican sector; even specific adjustments can be planned to identify the circumstances of the data collected. In addition, studies will be carried out on the technology of processing large amounts of data and information on cyber threats, as well as data mining analysis techniques, the analysis of correlation between the data with the main objective of processing the CTI information collected to estimate a forecast of cyber incidents based on evidence from intelligence sources.

References

1. W.E. Forum, «World Economic Forum Global Risks Perception Survey 2019–2020,» 2020. [En línea]. Available: http://www3.weforum.org/docs/WEF_Global_Risk_Report_2020.pdf
2. I. (. D. Corporation)., «The Growth in Connected IoT Devices Is Expected to Generate 79.4ZB of Data in 2025, According to a New IDC Forecast,» 2019. [En línea]. Available: https://www.idc.com/getdoc.jsp?containerId=prUS45213219
3. Gartner, «Leading the IoT: Gartner Insights on How to Lead in a Connected World.,» 2017. [En línea]. Available: https://www.gartner.com/imagesrv/books/iot/iotEbook_digital.pdf
4. Z. Doffman, «Cyberattacks on IoT Devices Surge 300% in 2019, 'Measured in Billions', Report Claims",» Forbes, 2019. [En línea]. Available: https://www.forbes.com/sites/zakdoffman/2019/09/14/dangerous-cyberattacks-on-iot-devices-up-300-in-2019-now-rampant-report-claims/#43ec06af5892
5. F-Secure, «ATTACK LANDSCAPE H1 2019,» 2019. [En línea]. Available: https://blog-assets.f-secure.com/wp-content/uploads/2019/09/12093807/2019_attack_landscape_report.pdf

6. A. Venkat, «Wikipedia Investigates DDoS Attack,» Bankinfosecurity.com, Information Security Media Group (ISMG), 2019 September. [En línea]. Available: https://www.bankinfosecurity.com/wikipedia-investigates-ddos-attack-a-13049
7. NIST, «COMPUTER SECURITY RESOURCE CENTER,» NIST, [En línea]. Available: https://csrc.nist.gov/glossary/term/cyber_risk. [Último acceso: 19 May 2020]
8. C. S. F. B. J. T. &. M. D. K. Kruse, «Cybersecurity in healthcare: A systematic review of modern threats and trend,» Technology and Health Care, 2017
9. D. R. A. Z. W. F. F. L. P. F. X. &. T. J. Wu, «Cybersecurity for digital manufacturing,» Journal of Manufacturing Systems, 2018
10. Fortinet, «Fortinet Threat Intelligence,» Fortinet, [En línea]. Available: https://www.fortinet.com/fortiguard/threat-intelligence/threat-research.html. [Último acceso: 03 May 2020]
11. Cisco, «Cisco Cybersecurity Report Series,» Cisco, [En línea]. Available: https://www.cisco.com/c/en/us/products/security/security-reports.html#~more-reports. [Último acceso: 03 May 2020]
12. Fireeye, «M-Trends 2020,» Fireeye, [En línea]. Available.: https://www.fireeye.com/current-threats/annual-threat-report/mtrends.html. [Último acceso: 03 May 2020]
13. ISO27000, « ISO/IEC 27032:2012 — Information technology — Security techniques — Guidelines for cybersecurity,» ISO/IEC, [En línea]. Available: https://www.iso27001security.com/html/27032.html. [Último acceso: 19 May 2020]
14. F. C. y. C. Teconologico. [En línea]. Available: http://foroconsultivo.org.mx/. [Último acceso: 08 May 2020]
15. I. a. d. i. y. t. d. México, «Foro Consultivo Cientifico y Tenologico,» 2018. [En línea]. Available: https://foroconsultivo.org.mx/proyectos_estrategicos/img/8/17.pdf. [Último acceso: 07 May 2020]
16. F. Staff, «Cibercrimen afecta a uno de cada cuatro mexicanos, según aseguradoras,» forbes Mexico, 05 May 2019. [En línea]. Available: https://www.forbes.com.mx/cibercrimen-afecta-a-uno-de-cada-cuatro-mexicanos-segun-aseguradoras/. [Último acceso: 08 May 2020]
17. E. s. C. E. e. M. 2. D. t. entrega, «Asociacion de Internet MX,» Diciembre 2019. [En línea]. Available: https://irp-cdn.multiscreensite.com/81280eda/files/uploaded/Estudio%20de%20Comercio%20Electro%CC%81nico%20en%20Me%CC%81xico%202019.pdf. [Último acceso: 08 May 2020]
18. P. d. c. e. México, «Comexi,» Junio 2018. [En línea]. Available: https://consejomexicano.org/multimedia/1528987628-817.pdf. [Último acceso: 08 May 2020]
19. G. d. Mexico, «Secretaria de Comunicaciones y Transportes,» [En línea]. Available: https://www.gob.mx/sct. [Último acceso: 12 May 2020]
20. OEA. [En línea]. Available: http://www.oas.org/es/. [Último acceso: 12 May 2020]
21. H. d. l. u. e. c. e. M. 2019, «Gobierno de Mexico,» 2019. [En línea]. Available: https://www.gob.mx/cms/uploads/attachment/file/444447/Estudio_Ciberseguridad.pdf. [Último acceso: 12 May 2020]
22. P. Federal, «¿Conoces qué es el Phishing?,» Gobierno de Mexico, 08 Enero 2019. [En línea]. Available: https://www.gob.mx/policiafederal/es/articulos/conoces-que-es-el-phishing?idiom=es. [Último acceso: 12 May 2020]
23. NIST, «NIST Information Technology Laboratory,» NIST, [En línea]. Available: https://www.nist.gov/itl/smallbusinesscyber/cybersecurity-basics/glossary. [Último acceso: 19 May 2020]
24. T. G. R. R. 2020, «The Global Risks Report 2020,» 16 January 2020. [En línea]. Available: http://www3.weforum.org/docs/WEF_Global_Risk_Report_2020.pdf. [Último acceso: 12 May 2020]
25. R. C. M. 2018, «Willis Towers Watson,» 04 December 2018. [En línea]. Available: https://www.willistowerswatson.com/es-MX/Insights/2018/12/riesgo-cibernetico-mexico-2018. [Último acceso: 13 May 2020]
26. J. Arreola, «Padrón electoral en la nube: ¿ciberproblemas a la mexicana?,» Forbes Mexico, 26 April 2016. [En línea]. Available: https://www.forbes.com.mx/padron-electoral-la-nube-ciberproblemas-la-mexicana/. [Último acceso: 13 May 2020]

27. A. México, «Comunicado Oficial: Sin afectaciones a datos o recursos de asegurados: AXA,» 23 Octubre 2018. [En línea]. Available: https://axa.mx/web/blog/postura-de-axa-mexico. [Último acceso: 19 May 2020]
28. N. Rial, «Mexican hackers attack official sites,» New Europe, 17 September 2012. [En línea]. Available: https://www.neweurope.eu/article/mexican-hackers-attack-official-sites/ . [Último acceso: 15 May 2020]
29. CONDUSEF, «FRAUDES CIBERNÉTICOS TRADICIONALES,» 2020. [En línea]. Available.: https://www.condusef.gob.mx/?p=estadisticas. [Último acceso: 18 May 2020]
30. Fortinet, «Threat Intelligence Insider Latin America,» Fortinet, 10 April 2020. [En línea]. Available: https://www.fortinetthreatinsiderlat.com/. [Último acceso: 20 May 2020]
31. E. n. d. Ciberseguridad, «Gobierno de Mexico,» 2017. [En línea]. Available: https://www.gob.mx/cms/uploads/attachment/file/271884/Estrategia_Nacional_Ciberseguridad.pdf. [Último acceso: 23 May 2020]
32. L. F. D. P. D. D. P. E. P. D. L. PARTICULARES, «http://www.diputados.gob.mx/» 2010. [En línea]. Available: http://www.diputados.gob.mx/LeyesBiblio/pdf/LFPDPPP.pdf. [Último acceso: 23 May 2020]
33. L. G. D. P. D. D. P. E. P. D. S. OBLIGADOS, «http://www.diputados.gob.mx/,» 2017. [En línea]. Available: http://www.diputados.gob.mx/LeyesBiblio/pdf/LGPDPPSO.pdf. [Último acceso: 23 May 2020]
34. G. d. Mexico, «¿Has sufrido acoso cibernético? ¡Identifica sus modalidades y protégete!,» [En línea]. Available: https://www.gob.mx/conavim/articulos/has-sufrido-acoso-cibernetico-te-decimos-a-donde-acudir. [Último acceso: 20 May 2020]
35. Excelsior, «Cómo denunciar delitos cibernéticos en México,» Excelsior, 05 May 2019. [En línea]. Available: https://www.excelsior.com.mx/hacker/como-denunciar-delitos-ciberneticos-en-mexico/1311256. [Último acceso: 20 May 2020]
36. S. d. Seguridad, «Unidad de Prevención e Investigación Cibernética,» Gobierno del Estado de México, [En línea]. Available: https://sseguridad.edomex.gob.mx/seguridad-publica-transito/policia-cibernetica. [Último acceso: 21 May 2020]
37. R. N. d. I. Financiera, «cnvb.gob.mx,» 2019. [En línea]. Available.: https://www.cnbv.gob.mx/Inclusi%C3%B3n/Documents/Reportes%20de%20IF/Reporte%20de%20Inclusion%20Financiera%209.pdf. [Último acceso: 25 May 2020]
38. C. N. B. y. d. Valores, «Foro de Ciberseguridad,» 2017, 2020 23 Oct. [En línea]. Available: https://www.gob.mx/cnbv/articulos/foro-de-ciberseguridad. [Último acceso: 25 May]
39. T. S. O. I. T. C. M. F. SYSTEM, «http://www.oas.org/,» 2019. [En línea]. Available: http://www.oas.org/en/sms/cicte/Documents/reports/The-State-of-Cybersecurity-in-the-Mexican-Financial-system.pdf. [Último acceso: 26 May 2020]
40. R. McMillan, «Definition: Threat Intelligence,» Gartner Research, 2016 May 2013. [En línea]. Available: https://www.gartner.com/en/documents/2487216/definition-threat-intelligence
41. 2. S. C. T. I. (. Survey, «https://www.sans.org/,» 2020. [En línea]. Available: https://www.sans.org/reading-room/whitepapers/analyst/2020-cyber-threat-intelligence-cti-survey-39395. [Último acceso: 27 May 2020]
42. Fortinet, «FortiGuard Labs,» Fortinet, [En línea]. Available: https://fortiguard.com/. [Último acceso: 01 06 2020]
43. C. systems, «Talos,» Cisco systems, [En línea]. Available: https://talosintelligence.com/. [Último acceso: 01 06 2020]
44. Fireeye, «Mandiant Threat Intelligence,» Fireeye, [En línea]. Available: https://www.fireeye.com/solutions/cyber-threat-intelligence.html. [Último acceso: 01 06 2020]
45. Facebook, «ThreatExchange Documentation,» Facebook, [En línea]. Available: https://developers.facebook.com/docs/threat-exchange/v2.12. [Último acceso: 06 01 2020]
46. P. K. D., V. M., A. J., T. F. Sudip Mittal$_*$, «CyberTwitter: Using Twitter to generate alerts for Cybersecurity Threats and Vulnerabilities,» 2016 IEEE/ACM International Conference on Advances in Social Networks Analysis and Mining (ASONAM), 2016
47. G. C. T. I. f. T. U. N. Classification, «Gathering Cyber Threat Intelligence from Twitter Using Novelty Classification,» 2019. [En línea]. Available: https://www.researchgate.net/publication/

334223932_Gathering_Cyber_Threat_Intelligence_from_Twitter_Using_Novelty_ Classification. [Último acceso: 01 06 2020]
48. Fortinet, «Fortinet Threat Intelligence Insider Latin America,» Fortinet, 27 11 2019. [En línea]. Available: https://www.fortinetthreatinsiderlat.com/es/Q1-2020/MX/html/trends#trends_position. [Último acceso: 01 06 2020]
49. Shodan, «What is Shodan?,» [En línea]. Available: https://help.shodan.io/the-basics/what-is-shodan. [Último acceso: 05 06 2020]
50. A. A. a. I. Alsmadi, «IoT and the Risk of Internet Exposure: Risk Assessment Using Shodan Queries,» 2019 IEEE 20th International Symposium on A World of Wireless, Mobile and Multimedia Networks, 2019
51. Shodan, «Mexico Internet Exposure Dashboard,» [En línea]. Available: https://exposure.shodan.io/#/MX. [Último acceso: 05 06 2020]
52. Shodan, «Shodan,» [En línea]. Available: https://www.shodan.io/search?query=country%3A%22MX%22. [Último acceso: 05 06 2020]
53. B. Packets®, «Bad Packets® Cyber Threat Intelligence,» [En línea]. Available: https://badpackets.net/threat-intelligence/. [Último acceso: 05 06 2020]
54. B. Packets®, «Mirai-like Botnet Hosts,» Bad Packets®, [En línea]. Available: https://mirai.badpackets.net/accounts/login/?next=/%3Fpage%3D27651%26sort%3Dcountry. [Último acceso: 05 06 2020]
55. A. K. M. V. G. a. T. M. O. P. Dwyer, «Profiling IoT-Based Botnet Traffic Using DNS,» 019 IEEE Global Communications Conference (GLOBECOM), 2019
56. V. G. a. T. M. Angelos K. Marnerides, «Identifying infected energy systems in the wild,» de e-Energy '19: Proceedings of the Tenth ACM International Conference on Future Energy Systems, Phoenix AZ, 2019
57. P. Feeds, «openphish,» [En línea]. Available: https://openphish.com/phishing_feeds.html. [Último acceso: 06 06 2020]
58. openphish, «https://openphish.com/feed.txt,» [En línea]. Available: https://openphish.com/feed.txt. [Último acceso: 06 06 2020]
59. URLhaus, [En línea]. Available: https://urlhaus.abuse.ch/feeds/country/MX/. [Último acceso: 07 06 2020]
60. N. Young, «github.com,» [En línea]. Available: https://github.com/twintproject/twint. [Último acceso: 07 06 2020]
61. Facebook, [En línea]. Available: https://developers.facebook.com/programs/threatexchange/. [Último acceso: 07 06 2020]
62. IBM, [En línea]. Available: https://www.ibm.com/mx-es/security/xforce. [Último acceso: 07 06 2020]
63. checkpoint, «Live Cyber Threat Map,» [En línea]. Available: https://threatmap.checkpoint.com/. [Último acceso: 07 06 2020]
64. A10, «DDOS WEAPONS INTELLIGENCE MAP,» [En línea]. Available: https://threats.a10networks.com/. [Último acceso: 07 06 2020]
65. mrlooquer, [En línea]. Available: https://mrlooquer.com/. [Último acceso: 08 06 2020]
66. M. Muñoz, M. Peralta, C.Y. Laporte, «nálisis de las debilidades que presentan las Entidades Muy Pequeñas al implementar el estándar ISO/IEC 29110: Una comparativa entre estado del arte y el estado de la práctica,» RISTI-Revista Ibérica de Sistemas e Tecnologias de Informação,, pp. 85–96, 2019
67. V. M. a. S. Bromander, «Cyber Threat Intelligence Model: An Evaluation of Taxonomies, Sharing Standards, and Ontologies within Cyber Threat Intelligence,,» 2017 European Intelligence and Security Informatics Conference (EISIC), 2017

Fusion of Digital Mammography with High-Resolution Breast PET: An Application to Breast Imaging

Liliana Reséndiz Sánchez, Luis Martin Torres Treviño, and Gisela Estrada Sánchez

Abstract One of the strategies to reduce mortality from breast cancer is based on screening using digital mammography as an initial evaluation. The detection and diagnosis of breast carcinomas are achieved by interpreting images of different modalities including digital mammograms, magnetic resonance imaging, ultrasound, and thermography; however, the literature shows that multimodal image fusion is highly accurate in representing breast carcinomas. Due to the human complexity in the diagnosis and detection of breast cancer and the implication of using historical patient imaging records, it is important to use processing tools that allow the analysis of breast images for possible improvements of the diagnosis. This chapter proposes the use of diversified data sets composed from different modalities to support the breast cancer diagnosis process and demonstrates that by applying various processing techniques it is possible to support the interpretation of the findings and that they can improve the precision in detecting breast cancer.

Keywords Breast imaging · Medical image fusion · Breast cancer imaging modalities · Diseases-based image fusion · Breast PET positron emission tomography · Fusion mammography with breast PET

L. R. Sánchez (✉)
Facultad de Ingeniería Mecánica y Eléctrica de la Universidad Autónoma de Nuevo León, Ciudad Universitaria, San Nicolás de los Garza, Nuevo León, Mexico
e-mail: liliana.resendizsnch@uanl.edu.mx

L. M. T. Treviño
Centro de Innovación, Investigación y Desarrollo en Ingeniería y Tecnología de la Universidad Autónoma de Nuevo León, Ciudad Apodaca, Mexico
e-mail: luis.torrestrv@uanl.edu.mx

G. E. Sánchez
CT Scanner del Sur, San Ángel, Álvaro Obregón, México
e-mail: dragiselaus@yahoo.com

F. Torres-Guerrero et al. (eds.), *2nd EAI International Conference on Smart Technology*, EAI/Springer Innovations in Communication and Computing,
https://doi.org/10.1007/978-3-031-07670-1_8

1 Introduction

Breast carcinoma could be effectively treated if caught early [3]. Therefore, it is important to have the right tools to notice the presence of signs of breast cancer. There are numerous tests and procedures for the prevention, diagnosis, and treatment, and one of the most important is digital mammography [11]. Clinically, digital mammography (MG) has been used as a standard test to diagnose breast cancer; this corresponds to a general examination and is useful for the detection of breast cancer and the reduction of mortality [6]. However, false positives on digital mammograms lead the second reviews, resulting in increased costs for their health care, as well as unnecessary medical procedures for patients [44]. Diagnostic ultrasound technique is recommended when breast density is reflected [45]. Given that small masses can pass through radiography radiation, the need to resort to other imaging modalities such as high-resolution breast PET could be more effective [16].

During the last two decades, molecular imaging has shown important advancement to integrate their techniques in the evaluation of malignant tumors. Equipment is specifically designed for breast examination and is characterized as a technique that offers greater spatial resolution that allows to detect smaller lesions. It is shown that the fusion of breast PET with mammography (PET/MG) imaging allows for more accurate evaluation by fusing anatomical location with functional imaging. The use of radiotracers in molecular imaging studies allows to detect breast carcinomas before vascularization since the metabolism of cancer cells generally increases before stimulation of the growth of new vessels [42]. The molecular image is obtained from the images from breast positron emission tomography, and it captures enough information to recognize possible oncological lesions at an early stage or not seen in the mammography that can be subject to quantitative evaluations for their detection, characterization, and monitoring.

In interest of improving lesion detection, the goal of this research is the use of diversified data sets of high-resolution breast PET with mammography images in a fused image to support the breast cancer imaging diagnostic process and demonstrate that by applying various processing techniques it is possible to correlate metabolic information to recognize important breast findings. The use of heterogeneous data sets is intended to provide support for a correct clinical diagnosis and can even perform the classification of features that allow the identification of the oncological lesion in malignant and benign groups through the selection, extraction, and classification of characteristics in the fused image.

Several recent studies are summarized in this chapter and indicate that high-resolution breast PET images combined with mammography images give enough evidence to be a useful diagnostic tool, although further evaluation and improvement may be required. So, we present the feasibility to analyze two types of heterogeneous data sets for clinical diagnostic purposes.

The remainder of this chapter is structured as follows: Sect. 2 gives some background of breast cancer screening; breast imaging technologies, mammography, and breast positron emission tomography are examined in Sect. 3. Fusion of mammography and high-resolution breast PET principles are inspected in Sect. 4. Material and methods are described in Sect. 5. Section 6 describes applications of deep learning in cancer detection, and finally, the conclusions are exposed in Sect. 7.

2 Breast Cancer Screening

Despite continued progress in detection and diagnosis, breast cancer is still an alarming global public health problem. Conventional mammography continues to be the cornerstone in the detection of breast carcinoma; however, new technologies provide valuable information on the molecular aspect of the tumor, with the consequent detection of small lesions, at earlier stages, with proper identification and better surgical planning, as well as decreased morbidity and mortality.

The premise of an early detection of breast cancer has a positive effect on the disease through medical intervention [33]. Through an early treatment, the reductions of the morbidity and mortality are still the main goals during significant finding detection.

3 Breast Imaging Technologies

3.1 Mammography

Through the analysis of mammograms, the presence of masses, calcifications, densities, among others, could be evaluated [13]. Several studies show lower sensitivity during the physical examination compared with the analysis of a mammogram [20]. When the goal is to detect calcifications, mammography is more accurate than ultrasound [13]. The diagnosis based on mammography has the ability to identify cancers due their absorption capability of x-rays with respect to the surrounding tissue [7], but there are high false negative and positive rates in patients with a dense breast tissue [46]. Also, mammography presents many drawbacks such as the use of ionizing radiation.

3.2 High-Resolution Breast PET

Historically, in the practice of nuclear medicine, medical specialists visually evaluate images for the detection and monitoring of breast carcinomas [10]. Although the expertise of a physician is considered as the most important factor during diagnosis, there are other aspects that definitely affect the final result, among them, image noise, the ability of visual perception from the physician, deficient image clarity, and inadequate contrast [23].

In 1994, Thompson et al. [50] developed a highly specific technique to detect the increased metabolic rate of breast tumors. The developed technique provides a low-cost, high-spatial-resolution positron imaging system known as high-resolution breast PET.

High-resolution breast PET uses a compression device that allows to detect 1.5 mm lesion [56]. Their technique compression device is solely for minimizing patient motion, is getting a more accurate result, and is also considered as an important tool for monitoring the cancer treatment response. Because cancerous cells present an increased glucose metabolism, the radiotracer molecules are taken up by the cells making suitable the localization of the cancer with high-resolution breast PET. Also in a PET/CT study, the metabolic activity in the tumor can be quantified to assist in assessing the effectiveness of therapy both during and after treatment, allowing for changes in treatment when needed [7].

It was demonstrated the potential to detect breast lesions with a series of special phantom experiments by measuring basic scanner parameters as scatter fraction, as well as sensitivity and major resolution [39]. Also, it was presented clinical results by analyzing images achieving a specificity over 90%, a sensitivity over 86%, and an accuracy of 89% in the diagnostic task that can be categorized as a feasible and accuracy rate [26]. High-resolution breast PET was considered as technique that can assist during the procedure of partial mastectomy to improve negative margins [48]. High-resolution breast PET showed higher accuracy results during the lesion characterization [56]. This imaging procedure is still considered as an emerging imaging technology that produces high-resolution tomographic 12-slice images of 18F-FDG uptake in the breasts [3].

Due to the advances in nuclear breast imaging devices, the interest in high-resolution breast PET has been increasing. Also because of the use of lower doses of radiopharmaceutical and their increased sensitivity, the high-resolution breast PET has been suggested for breast cancer detection and treatment planning [29].

Although, it is still considered a recently introduced nuclear medicine study, which after injecting a radiopharmaceutical called F-18 fluorodeoxyglucose (18-FDG) intravenously to subsequently acquire images of the mammary glands where it is possible to observe the behavior of lesions identified by other diagnostic modalities and their metabolism. Those suspicious breast lesions will have increased metabolism in this study. Silverstain et al. [45] reported that molecular imaging tools such as breast PET have equivalent sensitivity and improved specificity when it is compared with breast MRI so they recommended that breast PET is used when a contraindication is found for MRI in some patients. Specht et al. [47] characterized molecular imaging procedures that offer a better spatial resolution and greater accuracy when it comes to image quantification. Berg et al. [4] found greater specificity at the breast and lesion levels and show the performance when compared to MR imaging. A group of scientists identified the high-resolution breast PET as a useful tool due to the adequate results in early diagnosis of breast carcinoma [23], although further evaluation on improvement may be required. The evolution of positron emission tomography instruments and their requirements to obtain good-quality images are shown in the paper by Eo et al. [14].

High-resolution breast PET provides functional imaging information and is considered as a useful tool as a result of their high sensitivity because it can identify the stage of the breast cancer especially in those patients scheduled for conservative surgery as well as assess recurrence versus postsurgical changes and monitor the neoadjuvant chemotherapeutic response; the aforementioned contributes to improve the treatment planning of the disease [34].

It was concluded that the imaging sensitivity of high-resolution breast PET was higher than whole-body PET [57]. Also, there was found stronger correlations with immunohistochemical information of breast cancer using high-resolution breast PET up against to whole-body PET [32].

The values of specificity and sensitivity for breast PET images were established in different clinical situations by Martins MV et al. [30]. Molecular imaging is still considered as a worthy technique when it is combined with mammography [3, 42].

The effectiveness and characterization of breast images were studied, found a moderate positive predicted value, and considered the information of mammography should be used together to make diagnostic decision to improve the efficacy of studies [8].

The sensitivity and specificity as diagnostic values were evaluated by Farajati J et al. [18]; using the maximum high-resolution breast PET uptake value > 1.9, they concluded a specificity greater than 95% and with a sensitivity of 100%.

4 Fusion of High-Resolution Breast PET with Mammography

Efforts to demonstrate the benefits of using high-resolution breast PET with mammography have been different. Weinberg et al. [55] concluded in a device with biopsy capability combining with a conventional mammography; this allowed to exploit the potential of correlation of high-resolution breast PET with mammography. Thompson et al. [50] provide a low-cost imaging system in which high-resolution breast pet images can be correlated with mammography images.

Thompson [49] considered a group of designs as compatible when combining the high-resolution breast PET image with the mammography image. High diagnostic accuracy for breast lesions was found; when mammography and breast PET images were analyzed, it was found a sensitivity of 91%, a specificity of 93%, and an accuracy of 92%.

Fusion makes it possible to recognize breast cancer since the mammography image provides morphological information and the functional image provides metabolic information [17].

High-resolution breast PET contributed with extra information in patients with implants as well the hormonal status or even density does not affect the diagnose value. Also, breast PET imaging was very helpful to show breast cancer with more characterization, such as multifocal or multicenter disease and, in some cases, intraductal involvement [17].

The images provided by high resolution breast PET are suitable as a complementary study for detecting breast cancer [1].

Breast cancer cells show an increased absorption of the radiopharmaceutical than normal cells [17]. The fusion of breast PET with mammography was considered as an optimum choice due to their capability to provide additional morphological information in findings.

High-resolution breast PET and mammography are studies acquired in different conditions. Bergman et al. [5] proposed a process that makes simple the correlation of both modalities, and this allowed to obtain more accurate results during the registration of mammography image with functional image [5].

5 Material and Methods

The database was retrospectively reviewed for one hundred female patients with a suspicious breast lesion on mammography or clinical background. All breast PET images were reported by a nuclear medicine physician. The characterization of the images included breast density, right, left, or bilateral lesion, multifocality, multicentricity, and extension or intraductal component. Digital imaging and communications in medicine (DICOM) is used as the standard representation, communication, and storage of medical images and related information. A DICOM file format has been used, so we have implemented a tool for medical image registration that allows establishing correspondence between features in two sets of images, by using a rigid transformation model. The utilization of Grassroots DICOM library was chosen as a framework library.

5.1 Image Processing and Analysis

Image fusion combines information of two data sets of a related scenario, and this makes suitable to get additional information in a single scene. Image registration is a processing technique where two images are aligned by overlapping them; this allows to get a third integrated image. Both data sets were acquired with different conditions and devices. This task in which the input images are aligned before getting the fused image is called image registration.

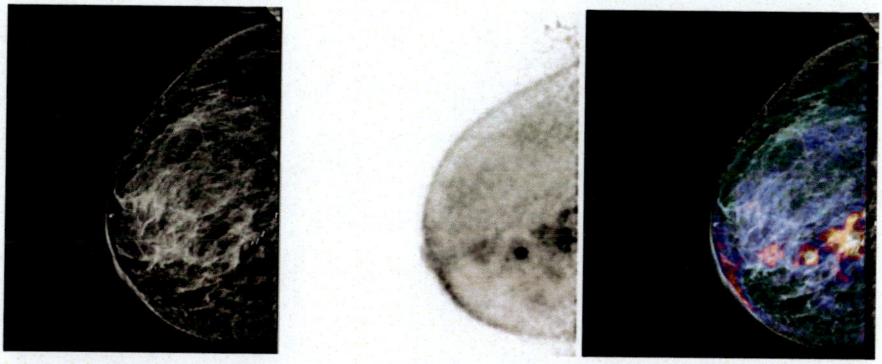

Fig. 1 Standard craniocaudal (CC) view of MG image. BIRADS 3 (left), high-resolution breast PET image (middle), and fused image previously registered (right). Patient with antecedent of ductal infiltrating carcinoma, treated with tumorectomy. Fusion PET/MG shows 18-FDG uptake near to the surgical staples and a multifocal lesion

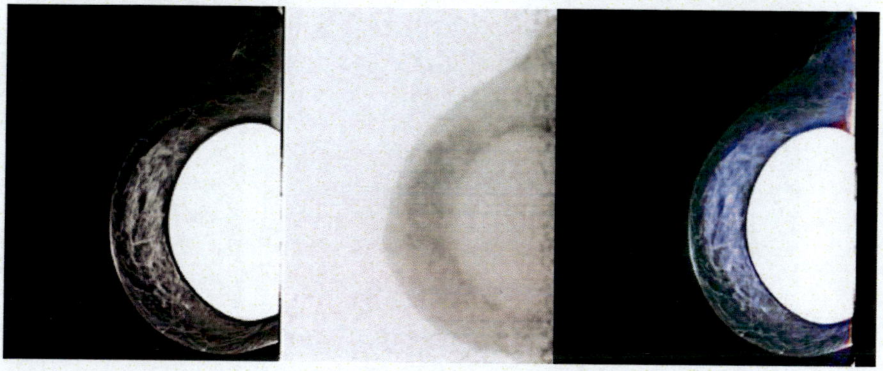

Fig. 2 Standard craniocaudal (CC) view of MG image (left), breast PET image (middle), and fused image previously registered (right), in a patient with breast implants

5.1.1 Image Registration

By finding the optimum geometrical transformation to correlate anatomical region between both images, valuable information is extracted and used to interpret and diagnose clinical findings that are considered as very complex tasks. One of the challenges of image registration is the aligning a series of two-dimensional images (Breast PET image) on a two dimensional image (MG image), due to different imaging condition in which different sensors are used to acquired to make a multimodal analysis as well low-quality and noise were present. The main goal is to integrate the information obtained from Breast PET device with a digital mammography image to get detailed scene representation. See Figs. 1, 2, 3, 4, and 5.

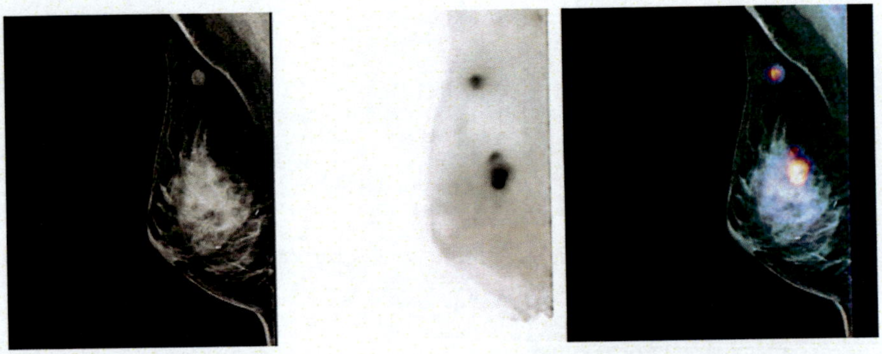

Fig. 3 The medio lateral oblique (MLO) view of MG image with density category: d (left), breast PET image (middle), and fused image previously registered (right). Patient with antecedent of a growing lymph node in the axilla, with two negative biopsies (surgical staples). Fusion PET/MG shows 18-FDG uptake in the upper quadrants, not suspected in the mammography

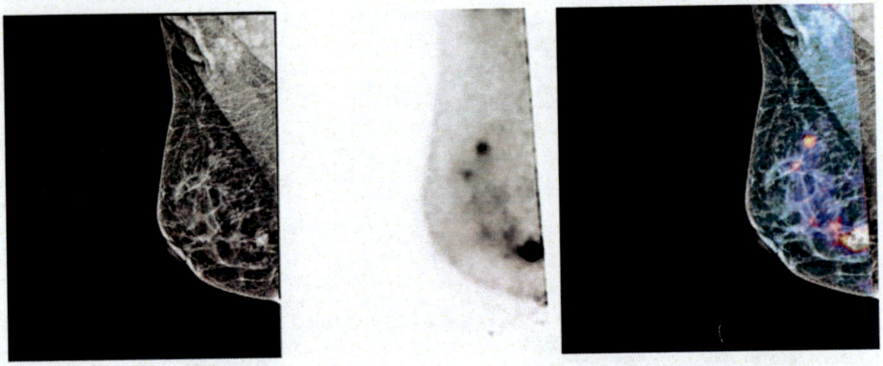

Fig. 4 The medio lateral oblique (MLO) view of MG image with one lesion in the inferior quadrant (left). High-resolution PET (middle) and fused PET/MG image (right) that show 18-FDG uptake in the same lesion shown in the mammography plus two other lesions in the upper quadrants (multicentric lesions)

In the last decade, several registration methods have considerably grown. One of the main contributions to image registration was described by Pluim et al. [36]. There are several applications of fusion between modalities such as computed tomography (CT) with positron emission tomography (PET), as well as magnetic resonance (MR) among others [24, 40, 41, 43, 54]. The image registration task is still facing new challenges and developments that will surely continue to position it as a very active area within image analysis and processing. Currently, there is a big initiative in the development of automatic and efficient registration techniques.

The images captured by breast PET are three-dimensional, and mammography images are considered a 2D imaging modality. The complexity is visible because there is no positional information between both modalities, which represents a challenge during the registration process. For the purpose of depicting fused image,

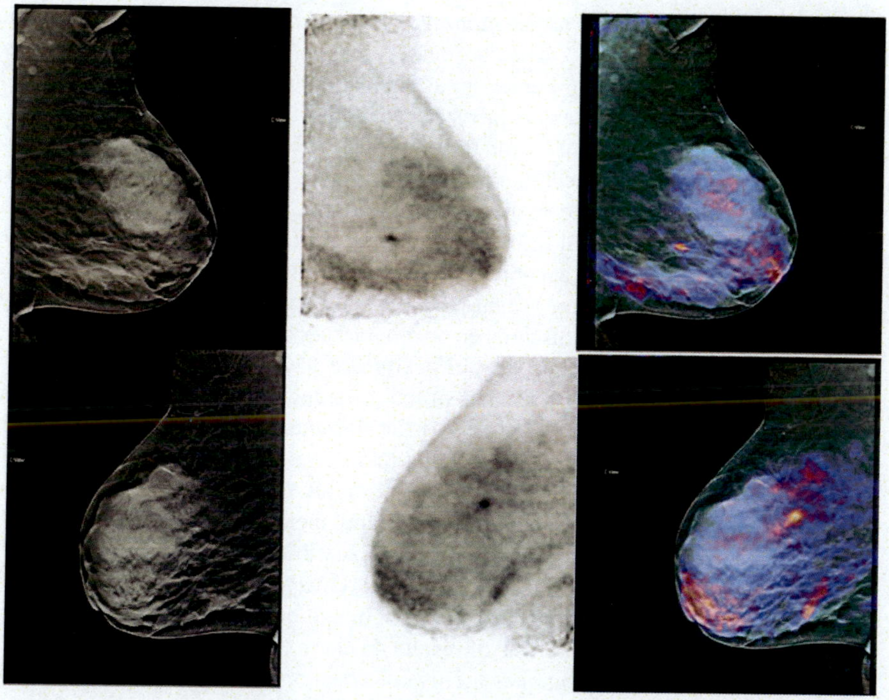

Fig. 5 The mediolateral oblique (MLO) view of MG image (left), high-resolution breast PET image (middle), and fused image previously registered (right). Left breast (upper images). Right breast (inferior images). Patient with mother and sister with breast cancer, BIRADS 3. Fusion PET/MG shows focal 18-FDG uptake in both breast. Bilaterality

we studied the registration methodologies based on intensity and their features. The methodologies included the common geometrical transformations and professional assessment techniques realized by physicians.

6 Applications of Deep Learning in Breast Cancer Detection

In the last two decades, there have been substantial advances in methods to detect breast cancer with artificial intelligence techniques. The complexity of each particular task makes the workflow meaningful. But with the development of new machine learning methods and their application in the clinical area, the need for precision is crucial to a major contribution to classification, diagnosis, and planning treatment [15]. Litjens et al. [28] evidence the existence of demand in the application of machine learning models for the purposes of prediction and prognosis of cancer due to a prominent need for personalized treatments.

Currently, there are applications assisted by Artificial Intelligence techniques that are already in use, and their constant evaluation of their performance is tracked to refine the ways of communicating to the patient that risk information, as well as to the doctor who provides the primary care [15].

The consideration of hundreds and sometimes thousands of clinical cases allows artificial intelligence techniques to recognize the subtle patterns of breast tissue that are considered precursors of breast carcinomas. The techniques allow learning by taking advantage of all information directly from the data by creating models that are significantly more accurate in various populations. The use of applications based on artificial intelligence techniques allows additional assistance, thus achieving a double diagnosis that discards errors on a larger scale [35].

Deep learning applications in healthcare have grown in importance in recent years [52], and the performance of the different techniques in object detection and classification tasks has been key for its use and application with medical images. Automatic feature extraction still presents a constant and ongoing challenge to find the features that accurately describe the output versus input data. The reproducible capacity of deep learning techniques and their non-discriminatory approach to characteristics makes the implementation feasible [38].

Dreyer and Allen [12] show the importance of using platforms to handle large amounts of information derived from image-based medical records, as well as emphasize effective analysis of results [12].

Deep learning techniques are useful when applied to various fields of research using diversified data sets collected from different sources [53], enhancing the diagnostic process in the medical area, which helps to spread the hypothesis through the application of various techniques that they allow to predict the acceleration of the multiple repetitive tasks of doctors [31].

An analysis of different studies that exploit deep architectures was carried out and is presented in the paper by Hamidinekoo et al. [19]. In this analysis, it has been identified the convolutional neural network as the most common architecture.

Arevalo et al. [2] tested several conventional neural network architectures and compared them with two descriptors during the manual diagnosis of injuries. Their experimentation was carried out with the BCDR-FM data set, and it was not tested with pre-trained networks.

The use of mammography images with a combination of pre-trained convolutional neural network is shown in the work realized by Carneiro et al. [9]; they found these models useful in medical applications and showed that it is not necessary a pre-registration of the input images in a multiview classification. Additionally, the risk of breast cancer is established according to BIRADS. As a result, the pre-trained models show better performance against the randomized ones. Huynh et al. [21] used the pre-trained AlexNet to address mass diagnosis by analyzing the performance of support vector machines (SVM) as a classifier. A scheme in which a convolutional neural network that has been pre-trained was adjusted in a subset of the DDSM database is presented by Jiao et al. [22]. The features that represent the masses were extracted from the layers of the model using different scales that correspond to high and medium levels, and then the use of support vector machine was used as a classifier merging their predictions in each case. Levy and Jain [27] adopt AlexNet and GoogleNet architecture to classify findings in mammography images. They explore transferred learning and compare against one made from scratch.

Ting et al. [51] proposed a deep classification algorithm using mammography images of MIAS database and built an architecture including 28 convolutional layers. Rampun et al. [37] adapted AlexNet architecture by modifying to get a new pre-trained version, then adjust with curated breast imaging subset of DDSM (CBIS-DDSM), and made their predictions based on three models. Lehman et al. [25] developed an algorithm and trained a deep convolutional neural network based on ResNet-18 architecture, to measure the amount of fibrous and glandular tissue. There is a major risk of breast cancer in women with dense breast as the tumors can be masked.

In summary, the technology can improve user practice through the use of artificial intelligence algorithms as an aid in the management of data science, tools, and knowledge in medicine to incorporate them into patient care. The literature review shows substantial efforts in the area of artificial intelligence applied to medicine and has addressed the improvement of algorithms as well as their accuracy, execution, and propagation in volumes of data as well as in the application in electronic records of the health. It is important to emphasize the veracity of the information that is used during the training phase as well as in the test phase to obtain greater accuracy in any diagnostic result.

7 Conclusions

Although there are several imaging options capable of identifying and defining breast cancer, the fusion of breast PET combined with mammography can provide additional information for the detection of the primary lesion since breast PET measures metabolism, mammography images offer anatomical reference through different views of each breast that can be evaluated together by the interpreting physicians, and merging both techniques allows the anatomical localization related to the functional image. The fused image can be obtained through the application of conventional image analysis and processing techniques as well as artificial intelligence techniques. Breast PET and MG are synergistic and when combined in a single image, allow to detect minor findings specially in patients with dense breast. The literature review shows the fusion of breast PET findings with mammography (PET/MG) allowing to identify the primary lesion in dense breast, multifocal disease, multicentric disease, bilaterality or ductal involvement. Integrating information between these images could increase the specificity, sensitivity, accuracy, and the positive predictive value of MG in diagnostic work-up of breast cancer.

References

1. L. Antunovic, F. Gallivanone, M. Sollini, A. Sagona, A. Invento, G. Manfrinato, M. Kirienko, C. Tinterri, A. Chiti, I. Castiglioni, [18f]FDG PET/CT features for the molecular characterization of primary breast tumors. Eur. J. Nucl. Med. Mol. Imag. **44**(12), 1945–1954 (2017). https://doi.org/10.1007/s00259-017-3770-9
2. J. Arevalo, F. González, R. Ramos, J. Oliveira, M.A. Guevara Lopez, Representation learning for mammography mass lesion classification with convolutional neural networks. Comput. Methods Programs Biomed. **127**, 248–257 (2015). https://doi.org/10.1016/j.cmpb.2015.12.014
3. W.A. Berg, I.N. Weinberg, D. Narayanan, M.E. Lobrano, E. Ross, L. Amodei, L. Tafra, L.P. Adler, J. Uddo, W. E.A.L. Stein, High-resolution fluorodeoxyglucose positron emission tomography with compression ("positron emission mammography") is highly accurate in depicting primary breast cancer. Breast J. **12**(4), 309–323 (2006). https://doi.org/10.1111/j.1075-122x.2006.00269.x
4. W.A. Berg, K.S. Madsen, K. Schilling, M. Tartar, E.D. Pisano, L.H. Larsen, D. Narayanan, A. Ozonoff, J.P. Miller, J.E. Kalinyak, Breast cancer: comparative effectiveness of positron emission mammography and MR imaging in presurgical planning for the ipsilateral breast. Radiology **258**(1), 59–72 (2011). https://doi.org/10.1148/radiol.10100454
5. A.M. Bergman, C.J. Thompson, K. Murthy, J.L. Robar, R.L. Clancy, M.J. English, A. Loutfi, R. Lisbona, J. Gagnon, Technique to obtain positron emission mammography images in registration with x-ray mammograms. Med. Phys. **25**(11), 2119–2129 (1998). https://doi.org/10.1118/1.598408
6. I. Bleiweiss, Pathology of invasive breast cancer, in *Breast Cancer* (Elsevier, Amsterdam, 2005), pp. 98–110. https://doi.org/10.1016/b978-0-443-06634-4.50012-1
7. K.L. Bontrager, *Bontrager's Textbook of Radiographic Positioning and Related Anatomy*, 9th edn. (Elsevier, St. Louis, 2017–2018)
8. A. Boonyaleepan, Positron emission mammography for breast cancer in Rajavithi Hospital. J. Med. Assoc. Thailand = Chotmaihet Thangphaet **99**(suppl 2), S130–5 (2016). http://europepmc.org/abstract/MED/27266227
9. G. Carneiro, J. Nascimento, A.P. Bradley, Unregistered multiview mammogram analysis with pre-trained deep learning models, in *Medical Image Computing and Computer-Assisted Intervention – MICCAI 2015*, ed. by N. Navab, J. Hornegger, W.M. Wells, A.F. Frangi (Springer, Cham, 2015), pp. 652–660
10. I.H. Cho, E.J. Kong, Potential clinical applications of 18f-fluorodeoxyglucose positron emission tomography/magnetic resonance mammography in breast cancer. Nucl. Med. Mol. Imag. **51**(3), 217–226 (2016). https://doi.org/10.1007/s13139-016-0446-5
11. M.D. Dorrius, E.F.J. de Vries, R.H.J.A. Slart, A.W.J.M. Glaudemans, Breast cancer: a new imaging approach as an addition to existing guidelines. Eur. J. Nucl. Med. Mol. Imag. **42**(6), 813–817 (2015). https://doi.org/10.1007/s00259-015-3032-7
12. K. Dreyer, B. Allen, Artificial intelligence in health care: brave new world or golden opportunity? J. Am. College Radiol. **15**(4), 655–657 (2018). https://doi.org/10.1016/j.jacr.2018.01.010
13. A.V. D'Amico, J.S. Loeffler, J.R. Harris, *Image-Guided Diagnosis and Treatment of Cancer* (Humana Press, Totowa, 2003). https://doi.org/10.1007/978-1-59259-422-1
14. J.S. Eo, I.K. Chun, J.C. Paeng, K.W. Kang, S.M. Lee, W. Han, D.Y. Noh, J.K. Chung, D.S. Lee, Imaging sensitivity of dedicated positron emission mammography in relation to tumor size. Breast **21**(1), 66–71 (2012). https://doi.org/10.1016/j.breast.2011.08.002
15. B.J. Erickson, P. Korfiatis, Z. Akkus, T.L. Kline, Machine learning for medical imaging. RadioGraphics **37**(2), 505–515 (2017). https://doi.org/10.1148/rg.2017160130
16. L.A. Espinosa, M.B.J. Bernal, G.E. Sánchez, R. de la Mora Cervantes, J.L.C. Cortés, M. del Carmen Lara Tamburrino, Mastografía por emisión de positrones: revisión de un promisorio instrumento diagnóstico. Gaceta Mexicana de Oncología **15**(2), 78–85 (2016). https://doi.org/10.1016/j.gamo.2016.03.004

17. G. Estrada-Sanchez, High resolution breast pet as an invaluable tool to diagnose new breast cancer lesions in patients with dense breast. J. Nucl. Med. **60**(suppl. 1), 1240 (2019)
18. J. Farahati, A.G. Müller, E. Gillman, M. Hentschel, F.H.H. Müller, Positron emission mammography in the diagnosis of breast cancer. Nuklearmedizin **55**(01), 15–20 (2016). https://doi.org/10.3413/nukmed-0753-15-07
19. A. Hamidinekoo, E. Denton, A. Rampun, K. Honnor, R. Zwiggelaar, Deep learning in mammography and breast histology, an overview and future trends. Med. Image Anal. **47**, 45–67 (2018). https://doi.org/10.1016/j.media.2018.03.006
20. M.A. Helvie, L.K. Joynt, R.L. Cody, L.J. Pierce, D.D. Adler, S.D. Merajver, Locally advanced breast carcinoma: accuracy of mammography versus clinical examination in the prediction of residual disease after chemotherapy. Radiology **198**(2), 327–332 (1996). https://doi.org/10.1148/radiology.198.2.8596826
21. B. Huynh, H. Li, M. Giger, Digital mammographic tumor classification using transfer learning from deep convolutional neural networks. J. Med. Imag. **3**, 034501 (2016). https://doi.org/10.1117/1.JMI.3.3.034501
22. Z. Jiao, X. Gao, J. Li, A deep feature based framework for breast masses classification. Neurocomputing **197**, 221–231 (2016). https://doi.org/10.1016/j.neucom.2016.02.060
23. V. Kalles, G.C. Zografos, X. Provatopoulou, D. Koulocheri, A. Gounaris, The current status of positron emission mammography in breast cancer diagnosis. Breast Cancer **20**(2), 123–130 (2012). https://doi.org/10.1007/s12282-012-0433-3
24. B. Karaçali, Information theoretic deformable registration using local image information. Int. J. Comput. Vis. **72**(3), 219–237 (2006). https://doi.org/10.1007/s11263-006-8704-0
25. C. Lehman, A. Yala, T. Schuster, B. Dontchos, M. Bahl, K. Swanson, R. Barzilay, Mammographic breast density assessment using deep learning: clinical implementation. Radiology **290**, 180694 (2018). https://doi.org/10.1148/radiol.2018180694
26. E.A. Levine, R.I. Freimanis, N.D. Perrier, K. Morton, N.M. Lesko, S. Bergman, K.R. Geisinger, R.C. Williams, C. Sharpe, V. Zavarzin, I.N. Weinberg, P.Y. Stepanov, D. Beylin, K. Lauckner, M. Doss, J. Lovelace, L.P. Adler, Positron emission mammography: Initial clinical results. Ann. Surgical Oncol. **10**(1), 86–91 (2003). https://doi.org/10.1245/aso.2003.03.047
27. D. Lévy, A. Jain, Breast mass classification from mammograms using deep convolutional neural networks. CoRR abs/1612.00542 (2016). http://arxiv.org/abs/1612.00542
28. G. Litjens, T. Kooi, B.E. Bejnordi, A.A.A. Setio, F. Ciompi, M. Ghafoorian, J.A. van der Laak, B. van Ginneken, C.I. Sánchez, A survey on deep learning in medical image analysis. Med. Image Anal. **42**, 60–88 (2017). https://doi.org/10.1016/j.media.2017.07.005
29. L. MacDonald, J. Edwards, T. Lewellen, D. Haseley, J. Rogers, P. Kinahan, Clinical imaging characteristics of the positron emission mammography camera: PEM Flex Solo II. J. Nucl. Med. **50**(10), 1666–1675 (2009). https://doi.org/10.2967/jnumed.109.064345
30. M.V. Martins, Positron emission mammography, in *Mammography Techniques and Review.* InTech (2015). https://doi.org/10.5772/60452
31. R.V. Milani, N.C. Franklin, The role of technology in healthy living medicine. Prog. Cardiovasc. Dis. **59**(5), 487–491 (2017). https://doi.org/10.1016/j.pcad.2017.02.001
32. A. Moscoso, Á. Ruibal, I. Domínguez-Prado, A. Fernández-Ferreiro, M. Herranz, L. Albaina, S. Argibay, J. Silva-Rodríguez, J. Pardo-Montero, P. Aguiar, Texture analysis of high-resolution dedicated breast 18 f-FDG PET images correlates with immunohistochemical factors and subtype of breast cancer. Eur. J. Nucl. Med. Mol. Imag. **45**(2), 196–206 (2017). https://doi.org/10.1007/s00259-017-3830-1
33. S.J. Nass, I.C. Henderson, J.C. Lashof, *Mammography and Beyond: Developing Technologies for the Early Detection of Breast Cancer* (The National Academies Press, Washington, 2001). https://doi.org/10.17226/10030,
34. M. Noritake, K. Narui, T. Kaneta, S. Sugae, K. Sakamaki, T. Inoue, T. Ishikawa, Evaluation of the response to breast cancer neoadjuvant chemotherapy using 18f-FDG positron emission mammography compared with whole-body 18f-FDG PET. Clin. Nucl. Med. **42**(3), 169–175 (2017). https://doi.org/10.1097/rlu.0000000000001497

35. F. Pesapane, M. Codari, F. Sardanelli, Artificial intelligence in medical imaging: threat or opportunity? Radiologists again at the forefront of innovation in medicine. Eur. Radiol. Exp. **2**(1), 35 (2018). https://doi.org/10.1186/s41747-018-0061-6
36. J. Pluim, J. Fitzpatrick, Image registration. IEEE Trans. Med. Imag. **22**(11), 1341–1343 (2003). https://doi.org/10.1109/tmi.2003.819272
37. A. Rampun, B. Scotney, P. Morrow, H. Wang, Breast mass classification in mammograms using ensemble convolutional neural networks, in *Breast Mass Classification in Mammograms Using Ensemble Convolutional Neural Networks* (2018), pp. 1–6. https://doi.org/10.1109/HealthCom.2018.8531154
38. D. Ravi, C. Wong, F. Deligianni, M. Berthelot, J. Andreu-Perez, B. Lo, G.Z. Yang, Deep learning for health informatics. IEEE J. Biomed. Health Inf. **21**(1), 4–21 (2017). https://doi.org/10.1109/jbhi.2016.2636665
39. R.R. Raylman, S. Majewski, R. Wojcik, A.G. Weisenberger, B. Kross, V. Popov, H.A. Bishop, The potential role of positron emission mammography for detection of breast cancer. A phantom study. Med. Phys. **27**(8), 1943–1954 (2000). https://doi.org/10.1118/1.1287439
40. T. Rohlfing, C. Maurer, D. Bluemke, M. Jacobs, Volume-preserving nonrigid registration of MR breast images using free-form deformation with an incompressibility constraint. IEEE Trans. Med. Imag. **22**(6), 730–741 (2003). https://doi.org/10.1109/tmi.2003.814791
41. D. Rueckert, L.I. Sonoda, C. Hayes, D.L.G. Hill, M.O. Leach, D.J. Hawkes, Nonrigid registration using free-form deformations: application to breast MR images. IEEE Trans. Med. Imag. **18**(8), 712–721 (1999)
42. K. Schilling, Positron emission mammography: better than magnetic resonance mammography? Eur. J. Radiol. **81**, S139–S141 (2012). https://doi.org/10.1016/s0720-048x(12)70058-x
43. J. Schnabel, C. Tanner, A. Castellano-Smith, A. Degenhard, M. Leach, D. Hose, D. Hill, D. Hawkes, Validation of nonrigid image registration using finite-element methods: application to breast MR images. IEEE Trans. Med. Imag. **22**(2), 238–247 (2003). https://doi.org/10.1109/tmi.2002.808367
44. N. Sharma, D. Neumann, R. Macklis, The impact of functional imaging on radiation medicine. Rad. Oncol. **3**(1), 25 (2008). https://doi.org/10.1186/1748-717x-3-25
45. M.J. Silverstein, M.D. Lagios, A. Recht, D.C. Allred, S.E. Harms, R. Holland, D.R. Holmes, L.L. Hughes, R.J. Jackman, T.B. Julian, H.M. Kuerer, H.C. Mabry, D.R. McCready, K.M. McMasters, D.L. Page, S.H. Parker, H.A. Pass, M. Pegram, E. Rubin, A.T. Stavros, D. Tripathy, F. Vicini, P.W. Whitworth, Image-detected breast cancer: state of the art diagnosis and treatment. J. Am. College Surgeons **201**(4), 586–597 (2005). https://doi.org/10.1016/j.jamcollsurg.2005.05.032
46. A.L. Siu, Screening for breast cancer: U.S. preventive services task force recommendation statement. Ann. Internal Med. **164**(4), 279 (2016). https://doi.org/10.7326/m15-2886
47. J.M. Specht, D.A. Mankoff, Advances in molecular imaging for breast cancer detection and characterization. Breast Cancer Res. **14**(2), 206 (2012). https://doi.org/10.1186/bcr3094
48. L. Tafra, Z. Cheng, J. Uddo, M.B. Lobrano, W. Stein, W.A. Berg, E. Levine, I.N. Weinberg, D. Narayanan, E. Ross, D. Beylin, S. Yarnall, R. Keen, K. Sawyer, J.V. Geffen, R.L. Freimanis, E. Staab, L.P. Adler, J. Lovelace, P. Shen, J. Stewart, S. Dolinsky, Pilot clinical trial of 18f-fluorodeoxyglucose positron-emission mammography in the surgical management of breast cancer. Am. J. Surg. **190**(4), 628–632 (2005). https://doi.org/10.1016/j.amjsurg.2005.06.029
49. C.J. Thompson, Instrumentation for positron emission mammography. PET Clin. **1**(1), 33–38 (2006). https://doi.org/10.1016/j.cpet.2005.09.004
50. C. Thompson, K. Murthy, Y. Picard, I. Weinberg, R. Mako, Positron emission mammography (PEM): a promising technique for detecting breast cancer, in *Proceedings of 1994 IEEE Nuclear Science Symposium - NSS'94*. (IEEE, Piscataway, 1994). https://doi.org/10.1109/nssmic.1994.474696
51. F. Ting, Y. Tan, K. Sim, Convolutional neural network improvement for breast cancer classification. Exp. Syst. Appl. **120**, 103–115 (2018). https://doi.org/10.1016/j.eswa.2018.11.008

52. D. Ueda, A. Shimazaki, Y. Miki, Technical and clinical overview of deep learning in radiology. Jpn. J. Radiol. **37**(1), 15–33 (2018). https://doi.org/10.1007/s11604-018-0795-3
53. S.A. Wartman, C.D. Combs, Medical education must move from the information age to the age of artificial intelligence. Acad. Med. **93**(8), 1107–1109 (2018). https://doi.org/10.1097/acm.0000000000002044
54. C. Washington, M. Miga, Modality independent elastography (MIE): a new approach to elasticity imaging. IEEE Trans. Med. Imag. **23**(9), 1117–1128 (2004). https://doi.org/10.1109/tmi.2004.830532
55. I. Weinberg, S. Majewski, A. Weisenberger, A. Markowitz, L. Aloj, L. Majewski, D. Danforth, J. Mulshine, K. Cowan, J. Zujewski, C. Chow, E. Jones, V. Chang, W. Berg, J. Frank, Preliminary results for positron emission mammography: real-time functional breast imaging in a conventional mammography gantry. Eur. J. Nucl. Med. **23**(7), 804–806 (1996). https://doi.org/10.1007/bf00843710
56. I.N. Weinberg, D. Beylin, V. Zavarzin, S. Yarnall, P.Y. Stepanov, E. Anashkin, D. Narayanan, S. Dolinsky, K. Lauckner, L.P. Adler, Positron emission mammography: high-resolution biochemical breast imaging. Technol. Cancer Res. Treat. **4**(1), 55–60 (2005). https://doi.org/10.1177/153303460500400108
57. Y. Yamamoto, Y. Ozawa, K. Kubouchi, S. Nakamura, Y. Nakajima, T. Inoue, Comparative analysis of imaging sensitivity of positron emission mammography and whole-body PET in relation to tumor size. Clin. Nucl. Med. **40**(1), 21–25 (2015). https://doi.org/10.1097/rlu.0000000000000617

A Model for the Control and Monitoring of Supply Chain Indicators

Loraine Sanchez-Jimenez, Tomás E. Salais-Fierro, and Jania A. Saucedo-Martínez

Abstract In the current competitive environment, companies are pushed to develop strategies to achieve operational excellence in pursuit of growth and profitability. A supply chain focuses primarily on reducing costs by optimizing its processes, achieving a service level that meets the required quality standards. Managing the success of the supply chain is considered an essential activity in any organization. Then, the effectiveness and efficiency of the supply chain can be determined through a performance measurement system focused especially on logistics processes. The proposal established in this research consists of a system that integrates auxiliary techniques in decision-making with the aim of establishing performance indicators within the supply logistics process. In addition, this system incorporates fuzzy logic in order to establish more realistic and robust metrics and with the ability to feed back indicators under uncertain environments or with a lack of information. The presented system is cyclical and adaptive, which includes techniques based on AHP, SCOR, and Fuzzy Logic, and they support the decision-maker in any environment, stage, or process of the supply chain by determining through projections if the objectives planted in the improvement plans have been achieved. Additionally, it identifies the attributes that impact on the supply chain and those that represent areas of opportunity to improve.

Keywords Performance · Measurement · Techniques · Supply chain · Indicators · Models

---

L. Sanchez-Jimenez (✉) · T. E. Salais-Fierro · J. A. Saucedo-Martínez
Universidad Autónoma de Nuevo León, Ciudad Universitaria, San Nicolás de los Garza, Nuevo León, México
e-mail: tomas.salaisfr@uanl.edu.mx; jania.saucedomrt@uanl.edu.mx

F. Torres-Guerrero et al. (eds.), *2nd EAI International Conference on Smart Technology*, EAI/Springer Innovations in Communication and Computing,
https://doi.org/10.1007/978-3-031-07670-1_9

1 Introduction

Delivery on time adding value to the customer is probably the most challenging goal in the supply chain as its complexity has increased over the past years given the competition in the market aiming to meet the customer expectations. Different functional activities are taken into account in a supply chain. These are performed along the chain and include technology, business processes, and people, and infrastructure delivered a finished good or service [1]. Therefore, supply chain management has focused on maintaining an effective organization of activities from supplier to end customer by searching for ways to reduce or eliminate risks.

In the long term, it also establishes types of cooperation between the different actors in its supply chain to avoid any type of disruption and thus achieve better products and services [3].

A performance measurement system contributes to achieving business objectives [4]. Its structure is composed of attributes that measure the supply chain effectiveness and efficiency [5]. Measuring the correct process at the right time is vital for an improved decision-making process.

A system focused on analyzing the performance of any type of operating environment and its respective elements and/or processes can be designed from different perspectives, which can range from identifying the WIP and its characteristics to ROI or identifying any type of operating excess or unnecessary processes [6]. After selecting the aspects to be verified, the next step is to control and measure the information, finally, to evaluate and obtain results for subsequent improvement.

The frequent shortcomings in performance measurement system could be summarized in the lack of connection between strategic objectives and metrics, the biased centralization in finance and the existence of conflicting measures [7], useless metrics, and a guidelines that fall short on development at different levels and processes [8], not having a clear vision to establish the appropriate level of action required, long, medium, or short terms. In some cases, companies use benchmarking unequivocally by comparing themselves to large companies that are very different from a logistics standpoint or companies that are not logistically similar [9].

When the economic factor is established, a mechanism that determines and analyzes the supply chain must be an essential part of its control process. For this reason, establishing the level of performance is defined as an evaluation process to quantitatively and qualitatively determine the operational functionality of any business [10]. The evaluation is carried out by means of metrics or logistical indicators related to various performance objectives.

It is essential to define the scope of the performance evaluation when making a measurement system, to be clear about what to measure, how to do it, and above all to know the priorities for measurement. Also, it is important to consider the resources available to carry out the practice successfully.

In the past years, researchers have developed multiple supply chain performance frameworks to measure different problems or business models [11], and most researchers compose the performance measurement system based on several criteria [6]:

- Balanced scorecard (learning and development, internal business process, customer, and finance)
- Metrics (quantitative and qualitative): Classic versus innovative measures
- Decision-making levels (long, medium, or short terms)
- KPIs elements (resources, products, and flexibility)
- KPIs-based (financial and operational)
- Metrics location for the supply chain coordination (from planning to deliver processes)

Lima-Junior and Carpinetti [10, 12] indicate that one of the most popular indicators is cost. However, to the latter, it can be considered that the response capacity falls within the group of indicators most used in the logistics area, as well as flexibility, sustainability, among others.

The proposal presented in this research work is described in the following sections that have been divided as follows: the following section (background) analyzes some of the most well-known analysis tools in the industrial area in order to identify measurement systems according to the industry scenario. After this, the methodology applied in this research based on the findings found in the literature is presented. Then, the proposed system is described as well as the expected result of its application. At the end of the document, a reflection is made on the findings of the proposal, and the possible actions to improve its application are analyzed. It is concluded that the contributions of this system are the indicators obtained with respect to the base model applicable to the procurement area of the organization. Second, it determined the relationship among KPIs found with the base model and presented an evaluation of the main indicator systems. That is, it can be determined if the output projects adequate results and also evaluates the structure of measurement systems (the number of metrics, goals, relation to strategic objectives, etc.). Finally, configurations based on a hybrid model have a cyclic and adaptive system. This model is built from the metrics and attributes of the SCOR, a FAHP for the analysis of priorities and a FIS to measure predictively and defining the contribution of the factors with the greatest impact, with which a better alignment of the actions generated is achieved of the decision process and improving the results of the improvement plans.

2 Background

Some tools for the measurement and usage are revised in the following literature review.

Approaches	Process-based approaches	
	Perspective-based approaches	Supply Chain Operations Reference Model (SCOR)
		Balanced Scorecard (BSC)
	Hierarchy-based approaches	
Techniques	AHP - Analytical Hierarchy Analysis	
	Simulation	
	DEA – Análisis envolvente de datos	
Models	Deterministic Models	
	Stochastic Models:	
	Business Models	Multiple-criteria decision analysis (MCDA)

Fig. 1 Approaches, techniques, and models for performance measurement [13, 14]

2.1 *Techniques and Performance Evaluation*

System indicators drive performance. An incorrect assessment directly impacts the core operations of any business, resulting in lost revenue leading to poor growth year after year. So, implementing techniques for the implementation of supply chain evaluations is vital for its proper functioning.

[13, 14] show some examples of different methods of evaluating the good functioning of the systems. Figure 1 shows the summarized classification of the most commonly used tools over the years.

2.1.1 Approaches

Some classifications that have been found in the literature are the approaches based on processes, perspectives, and hierarchies. For this, the trends found in the area of performance measurement are established, and these are related to each other [13].

The first one supports the fulfillment of the organizations mission, orienting the necessary activities toward the satisfaction of the main actors of the chain: customers, suppliers, employees, shareholders, and consumer market in general [5]. The perspective-based model was developed considering that a particular set of objectives follows perspectives that consequently lead to a set of performance measure [15]. This model established a differentiated framework of metrics defined by understandable perspectives, in which there is an own interpretation of the challenges and solutions presented in the business and the metrics to be used. Through this approach, a general form of performance metrics is determined along with determining the correlation between the different metrics. Two main perspective models have been considered: models based on the balanced scorecard and models based on supply chain operation references [5]:

- Balanced Scorecard (BSC): It provides detailed information of the critical supply chain elements. It is an executive information system that monitors performance by relating strategies to objectives measured through indicators linked to action plans. In its structure, it takes into account five aspects: customers, internal business perspective, finance, and training and growth [7]. It is usual for companies using dashboards to be based on financial indicators.
- Supply Chain Operations Reference Model (SCOR): It is the model considered as the based model because of its wide use; its structure allows to join links of logistic processes, re-engineering, performance indicators, benchmarking, best practices, and technologies within the supply chain, which gives an improvement in the management and in the relationship of the actors [16]. It identifies the processes that are considered the main ones, which it groups into 5 actions that range from planning to return. These integrated processes provide a clear end-to-end perspective of the process and support optimizations internally and externally. It is used to describe any kind of supply chains using usual concepts.

When it comes to analyzing different hierarchical levels within a supply chain, hierarchy-based approaches should be considered for application. With this, it is possible to get the decision-maker to take an appropriate action according to the challenge that arises [5].

2.1.2 Techniques

In this type of systems, the techniques adopted for evaluation and development are those that are taken into account for their classification; among the most used are AHP, simulation, DEA, and fuzzy logic [5, 12, 13]:

- Analytical Hierarchical Process (AHP): It classifies the alternatives of a problem by deriving priorities. Multiple objectives are set from a decision-making problem. This hierarchy is supported by criteria, sub-criteria, and alternative decisions. The AHP works by comparing one criterion to the rest, measures the importance, although relative, of the element at different levels, and helps in the decision-making process [17].
- Data Envelopment Analysis (DEA): It is usually used in the analysis of benefits and costs when there are several decision criteria. It is a method based on mathematical techniques [18].
- Simulation: It is a technique that supports decision-making since it provides an approximation of a behavior that can occur under certain conditions that affect the system. Its use serves to reduce risks and costs (that would be incurred) by not making the right decision. Simulation facilitates the management of a supply chain; it creates a model of a complex system and includes random variables. Its main advantage is in the different applications, mainly allowing to study how the

whole system is affected by small and large changes without having to actually implement them [19].

2.1.3 Models

Generally, objectives and entities are established that are considered the feeders of the systems. In general terms, they are generated for the analysis and design of the supply chain: deterministic, stochastic, and business models. Among the differences in the models, they are considered deterministic in which the behavior of each of their components (variables) is known in advance. On the other hand, probabilistic (stochastic) models have a certain degree of uncertainty in one or more variables, although probabilistic distributions of their behavior can be established. Finally, business models integrate multiple and different performance indicators in a measurement system. The best known by their degree of use and importance according to the state of the art are BSC, SCOR, and DEA (defined above), and additionally, the multi-criteria analysis methodology (MCDA) is included [14]:

- Multiple-criteria decision analysis (MCDA): It describes all the processes that evaluate quantitative and qualitative elements simultaneously, considering factors that can be conflicting or decision analysis processes that involve two or more attributes. The overall objective of MCDA is to facilitate the choice of the correct decision when there is a series of alternatives in an environment of conflicting and competing positions. Several methods have been proposed in recent years to address problems of MCDA: Value function methods, objectives, and benchmarking methods and ranking methods [20].

2.2 *Approaches and Techniques Analysis*

According to what is established in Sect. 2.1, a research focused on hybrid tools or models used to measure multiple aspects in companies is carried out to confirm the findings or add new forms of measurement with the purpose of creating a hybrid model that can deliver a new input to the measurement systems and measure performance. The analysis included 23 articles classified by author, economic line of business of the company applying the measurement system, artificial intelligence techniques, and models. Companies from different industry areas were taken into account for this study, such as agricultural industries, transportation, even service companies, among others. Furthermore, the strategies measured mostly relate to sustainability, three-tier supply chains, customer perceived value, supplier selection, and evaluation. Table 1 lists the articles considered. Bold values at the bottom of the table (last row) indicate the total number of articles that uses each technique.

In Fig. 2 the summarized information found is presented. The measurements with the highest application are AHP. However, an analysis of the application of the tools

Table 1 Literature review

N	Author	Industry	Techniques							Models		IA		
			AHP	ANP	DEA	DELPHI	DEMATEL	Simulation	SEM	BSC	SCOR	Fuzzy	RNA	ANFIS
1	Sellitto et al. [41]	Footwear	x								x			
2	Tajbakhsh y Hassini (2015)	Manufacturer			x									
3	Bukhori et al.(2015)	Agriculture	x								x			
4	Tavana et al.(2016)	3-level SC			x									
5	Wibowo y Sholeh(2016)	Construction	x								x			
6	Yu et al.(2016)	Transportation			x									
7	Tavana et al.(2016)	Suppliers	x											x
8	Haghighi et al.(2016)	Recycling			x					x				
9	Brandenburg (2017)	Sales	x					x						
10	Govindan et al.(2017)	Manufacturing	x									x		
11	Singh et al. (2018)	Manufacturing	x							x		x		
12	Ramezankhani et al.(2018)	Manufacturing			x		x							
13	Thanki y Thakkar(2018)	Textile case		x			x			x		x		
14	Dissanayake y Cross (2018)	Manufacturing	x		x				x		x			
15	Rasolofo-Distler and Distler(2018)	Services								x				
16	Akkawuttiwanich and Yenradee [42]	Manufacturing									x	x		
17	Miranda et al.(2019)	Maintenance						x						
18	Lima-Junior y Ribeiro Carpinetti (2019)	Investigation									x		x	
19	Zanon et al.(2020)	Customer									x	x		
20	Chand et al.(2020)	Manufacturing				x	x							
21	Jollembeck Lopes y I. Pires (2020)	Galvanization		x								x		
22	Jiang et al. (2020)	Sustainability	x									x		
23	Lima-Junior y Carpinetti (2020)	Illustrative case									x			x
Total			**9**	**2**	**6**	**1**	**3**	**2**	**1**	**4**	**8**	**7**	**1**	**2**

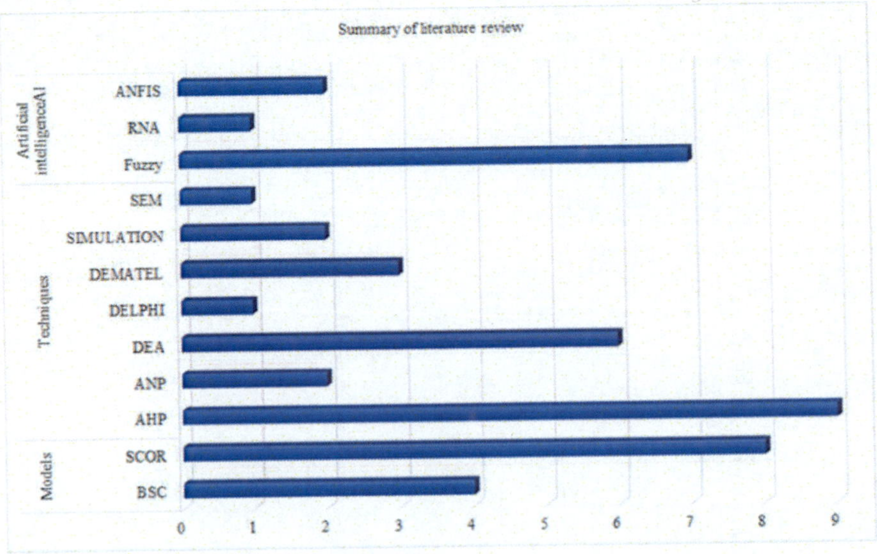

Fig. 2 Summary of literature review

is carried out with the two most widely used tools to identify their characteristics and functionalities and to choose the best option for the model.

As can be seen in the results, the metrics generated by the SCOR model are currently among the most popular, with the BSC model moving into second place. The SCOR model has a long record in research and case studies in many sectors throughout the years. This model is adaptive to the dynamic requirements of the user. Apart from the fact that the metrics considered optimal are established, the attributes represent different scenarios observed in the supply chains. It also enables a thorough assessment of the supply chain [21]. The main use of the score card model is mainly in finance despite its limitations such as fewer measures. Measuring the supply chain becomes a complex task by using this approach. In Table 2, the SCOR and BSC models are compared in terms of functionalities or the most relevant aspects, and it can be inferred that the representative figure for SCOR is due to its great adaptability.

With respect to techniques, AHP continues to be the most widely used followed by DEA. The first one is adopted for its wide applicability, simplicity, and great flexibility, in addition. It can be integrated with soft-computing techniques such as neural networks and fuzzy logic, to name a few, in order to create more robust hybrid methods. For the AHP, there are three principal operations: building hierarchies, a thorough analysis of priorities, and the verification of consistency. It is applied to determine costs and benefits, make planning, prioritize, etc. [22]. The second technique, DEA, in the allocation of resources is widely accepted, since it helps to establish qualitative analyses, in addition to helping to recognize inefficiencies and their origins, and any other evaluation that requires a qualitative approach [23].

Table 2 SCOR vs. BSC

SCOR	BSC
It generates a management system that transforms the strategy into tangible objectives and indicators.	Unifies terms and gives a standard format to describe the supply chain.
Prioritizes the most decisive processes for the success of the organization	Evaluates each process with indicators (KPIs) appropriate
Measures the impact of strategic decisions to check whether the allocation of resources of the organization is being effective	Maintains a continuous system of evaluation of KPIs and proposes future improvements

Hybrid models are techniques that are combined, which are currently being adopted. For the conformation of this model, methods that could work together with the SCOR model were investigated [24]. Making the right decision in an automated basis is a necessity addressed by the implementation of Artificial Intelligence that is lately being used in case investigations, illustrative tests, and case studies, to estimate the performance of a supply chain with different measures, and to forecast and check outcomes.

Fuzzy logic is a technique that manages vague data and knowledge, making it a useful approach when information is not available or when decision-makers.

AI provides different advantages that go beyond the adoption of guiding metrics [25]:the capacity to manage qualitative data and unpredictability for the decision-making process [26]; they are applicable and adaptable to the indicators established in the evaluation [25], and they are friendly to the operating scenario [10].

Finally, FST is added to manage unpredictability in the assessment process [27]. Therefore, to cover the lack of relevant information or inaccuracy in the data within the supply chain, a system that incorporates fuzzy logic is applied [28].

In order to propose a method that evaluates and feeds back the current measurement systems used in the companies and their performance in a cyclical way and in any layout of the supply chain, after analyzing the findings presented in this section, performance measures developed with the SCOR model in combination with fuzzy logic, FAHP, and FIS, a hybrid methodology is designed and presented. All this with the purpose of improving the actions and their respective results derived from a decision-making process.

3 Methodology

The methodology divided into the following three segments: [29]: a literature review, a thorough development, and the application. The changes made are set out in the supply chain configuration at the sourcing stage based on the SCOR performance attribute assumptions centralized in the same field of study. The components of the proposed approach are shown in Fig. 3.

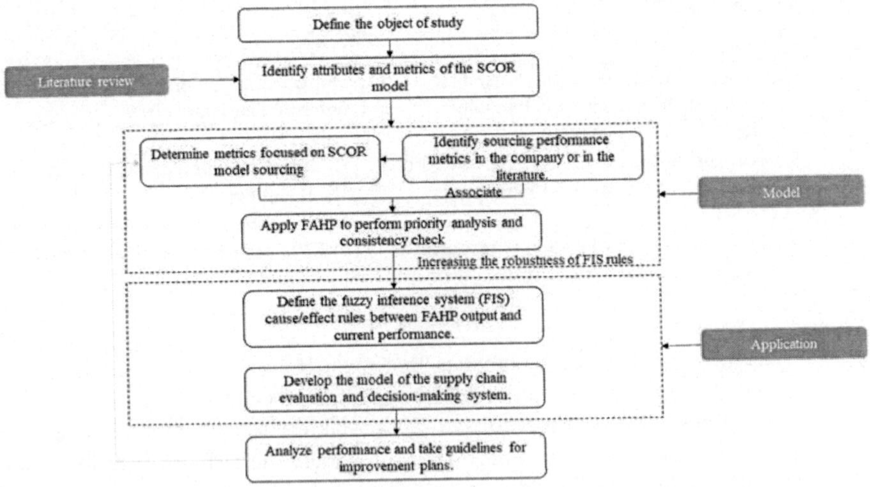

Fig. 3 Methodology

The first step would be a thorough review of the literature procuring the performance measurement. Also the SCOR model centered on indicators and attributes; as a priority analysis, we find the FAHP technique, and the fuzzy inference system (FIS) and its application to measure a supply chain.

As a second part of the approach, fuzzy logic is integrated into the AHP methodology, thereby creating a fuzzy AHP. Within this part of the approach, the results are extracted to generate the inference rules that will be processed in the FIS, which are a description of the trade-off and/or effects of the different performance metrics selected in the SCOR.

Finally, in order to demonstrate, evaluate, and validate the proposed method, it is important to acquire information generated in an actual case. For this, information generated in a company that provides its established indicators is captured, in addition to the incorporation of the user's experience.

For data collection, the model requires two inputs: the current performance of the purchasing-focused SCOR indicators, information collected through the company ERP system or metric indices, and the second input comes from user experience, which corroborates the singularities of the company environment: calculated through historical and linguistic scores. This information is applied in the construction of FIS rules.

The impact generated in this proposal is projected to focus on two main objectives, performance, and improvement. In other words, on the one hand, the sourcing area is analyzed and evaluated, and on the other hand, guidelines are obtained to develop improvement plans. In addition, the model is considered as a system component of continuous improvement, cyclical and receptive to any eventuality in the process. It is suggested that the validation of the information is to be carried out through a sensitivity analysis to verify the rules, the fuzzy operations, and the performance of the system. And, verify that the proposal is applicable in other categories of the supply chain (adaptive).

4 Development of the Methodology

4.1 Theoretical Constructs

- Sourcing: As its main function, the sourcing department is to ensure the quality of the materials and services provided by its suppliers. Today, managers emphasize evaluating the execution of this link by measuring and evaluating its contributions, delivering a positive result due to the ability to maximize value and minimize waste by taking proactive actions to improve efficiency and effectiveness in the chain [30].
- SCOR model attributes and metrics: The model establishes the link among people, processes, and best practices to meet the necessities of the end user with excellence in the supply chain [16]. SCOR recognizes five performance attributes, the first three defined below are considered client-centric, and the last two are internally focused:

  - Reliability: Ability to perform tasks as required, predictability. It takes into account delivery factors such as time, quantity, and adequate quality.
  - Responsiveness: Speed of the supply chain to perform tasks.
  - Agility: Ability to respond to changes due to external factors.
  - Cost: Operational costs of the supply chain.
  - Asset management: Ensures the organizational effectiveness with which resources and assets are used in order to meet demand.

  Level 1 strategic metrics are connected to the attribute that calculates if an organization thrives the desired positioning within the market space. Diagnostic metrics are recognized at three predefined levels. Table 3 relates the level 1 strategic metrics to their corresponding performance attributes in the SCOR model.
- Fuzzy AHP: It is based on three basic principles: the construction of a hierarchy, the comparative judgment of criteria using fuzzy numbers, and the synthesis of priorities [31–34]. The final result provides numerical priorities for the elements that embody the relative ability to achieve the objective [35]. This research proposes to use this technique to prioritize sourcing attributes and indicators to improve supply chain performance.
- Fuzzy inference system: It has been widely applied for multi-criteria decision-making due to its ability to deal with uncertainty [36, 37]. Also, it models human reasoning by means of fuzzy rules if-then [38, 39]. Furthermore, the application

Table 3 Level 1 metrics and performance attributes of the SCOR model

Level 1 strategic metrics	Performance attribute				
	External focus			Internal focus	
	Reliability	Responsiveness	Agility	Cost	Asset management
Perfect order fulfillment	■				
Order execution cycle time		■			
Superior supply chain adaptability			■		
Adaptability of the downstream supply chain			■		
Total value at risk			■		
Total supply chain management costs				■	
Costs of products sold				■	
Duration of the cash cycle					■
Performance of fixed assets in the supply chain					■
Return on working capital					■

Grey shaders indicate correspondence between the metric and the attributes.

of FIS in this context is appropriate because it allows to handle the non-linear relationship between input and output variables.

4.2 Model

The proposed model is shown in Fig. 4, illustrating the effects of the supply chain indicators in the sourcing area. Also showing the steps that this revolving structure makes up, and what and how they are part of a continuous improvement, allowing to have a comprehensive scheme of its performance taking into account all areas of the organization. On the other hand, including a simulation process allows detecting critical elements in the evolution and execution of the supply chain process. The steps of the presented approach are described below:

1. Identification of indicators focused on sourcing: As a first step, a set of SCOR metrics is defined: in this step, it is determined which attributes and metrics to use according to the interest; in this chapter, a search and compilation of

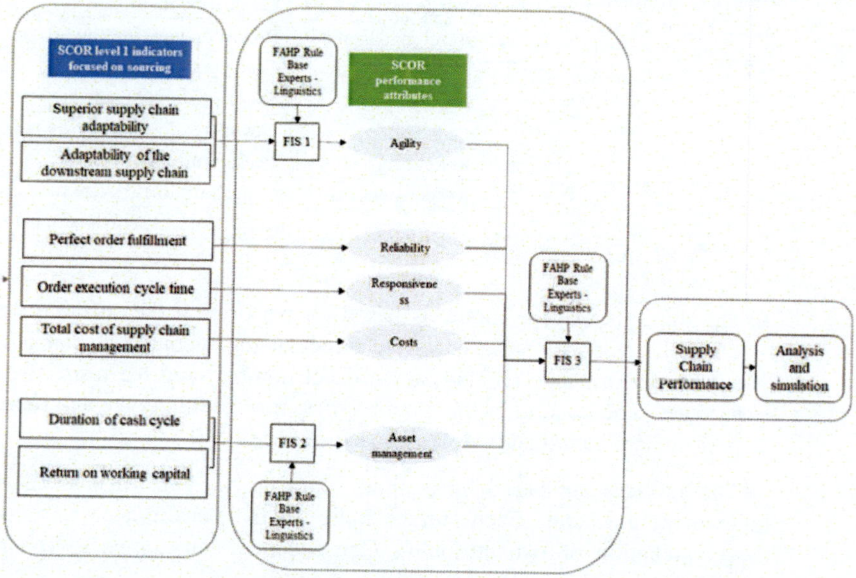

Fig. 4 Proposed model [29] .

information on the indicators focused on the SCOR model supply aligned with the competitive strategies to manage the supply chain is performed. The supply chain council [16] proposes the choice of at least one metric associated with each performance attribute in order to drive a balanced assessment and decision-making process based on different perspectives. Based on the review, all five attributes are considered appropriate for the purpose of the project. The ideal indicators identified in the hierarchical structure proposed by the SCOR model for this project are included in Table 4. It must be noted that depending on the supply chain measurement category, the attributes and metrics may vary, i.e., not necessarily all of them need to be included.

2. Fuzzy inference: The second step is to infer the sourcing values as a result of the performance of the SCOR level 1 indicators considered in Table 4. FIS 1 computes the agility from the comparable indicators. The parameters of this first FIS are rule-based and membership functions based on the perception of the panel of experts in supply chain and the performed process by the FAHP; for FIS 2, the same procedure is performed for the asset management attribute with its indicators. Finally, FIS 3 calculates the value of supply in five inputs: agility; the consequence of FIS 1, asset management; the consequence of FIS 2, perfect order fulfillment, order execution cycle time, and the total cost of supply chain management, the level 1 indicators of reliability, responsiveness and costs, respectively. It is in this stage where the semantic and the quantitative fuzzy data of the input and output variables are defined. Trilateral fuzzy numbers are frequently selected for these applications.

Table 4 Performance attributes and level 1 indicators

Attribute	Level 1 indicator
Asset management	Return on working capital
	Cash cycle time
Agility	Superior supply chain adaptability
	Downstream supply chain adaptability
Reliability	Perfect order fulfillment
Responsiveness	Order fulfillment cycle time
Costs	Total cost of supply chain management

3. The third step of the model is based on performing the scenario simulations using the response surfaces as the effect of the second FIS that provides the supply performance ranking, as an exact number, and the respective response surfaces. The ranking of the output varies in a specific range and represents the performance rating and the actual performance.

The surfaces are based on a figure showing the performance behavior as a function of attributes. Each surface indicates the performance as a function of the combination of two attributes. Consequently, ten comparative surfaces are generated:

- Reliability versus agility
- Reliability versus costs
- Reliability versus asset management
- Reliability versus responsiveness
- Agility versus responsiveness
- Agility versus costs
- Agility versus asset management
- Responsiveness versus costs
- Responsiveness versus asset management
- Costs vs. asset management

Surfaces are generated according to the number of attributes selected. The result of the simulation visualizes which attribute has the greatest impact on sourcing and which one has shortcomings; in this way, it is possible to make the required changes in the model and in the processes carried out in sourcing in order to obtain better results in performance. Additionally, with the results, it will be possible to make decisions of improvement.

4.3 Results

With the first step, the association of the company's indicators with those of the SCOR model is achieved. This part is necessary as this model is a benchmark and standard and can therefore be adapted to any metrics system used. In addition, using

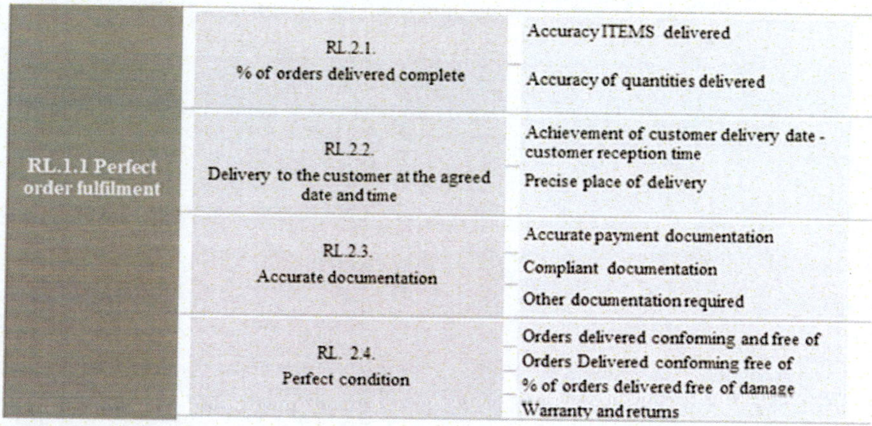

Fig. 5 Metrics levels 1, 2, and 3 of the reliability attribute

its language allows benchmarking that helps supply chains to identify gaps and make improvements. Indicators that range from levels 1 to 3 are considered relevant; therefore, it is advisable to pay attention to them. Since usually, companies do not consider the scope of their indicators and only focus on the local scope (specific to an area or a department). In contrast, the structures of level 1 and 2 indicators of the supply chain operations reference model are somewhat general. Therefore, it is essential to be familiar with all three levels. Figure 5 exemplifies the proposed revision method of the measurement levels covered by the reliability attribute.

Table 5 illustrates the relationship of KPIs and conversion of figures. Where the proportional relationship can be direct and inverse, it depends on the nature of the indicator [29]. For instance, there is better performance in the case of a higher direct ratio. The corresponding figures are converted to a range from 0 to 10 to make future comparative measurements possible. Equations (1) and (2) are used to find the values, respectively.

$$Converted\ figure = \frac{Current\ Figure}{Reference\ Figure} \tag{1}$$

$$Converted\ figure = \frac{Reference\ Figure}{Current\ Figure} \tag{2}$$

The FAHP methodology finds the priorities among the studied elements, i.e., for this case, according to the experts' conception, a categorization of the performance attributes from the most important to the least important is achieved by the value of their weights. These results, together with the conversion of the above figures, are the basis for the FIS rules.

Table 5 Association of indicators and conversion of figures

Indicator name literature	SCOR level 1 indicator	Unit	Current figure	Reference figure	Proportional relationship	Converted figure (range 0–10)
Compliance rate	Perfect order fulfilment	Percentage	0.69	0.88	Direct	8
Purchase order cycle time	Order execution cycle time	Time	7	5	Inverse	7

By applying the proposed FIS conceptual model, the result projects ten comparison surfaces between the attributes of the SCOR model: agility, responsiveness, cost and asset management. These figures express a scenario analysis of sourcing behavior under certain parameters and provide decision-makers with information on the impact on the supply chain. Based on the behavior observed through FIS 3, improvement plans can be designed connected to the organizations projected goals to optimize the current state.

The results of the simulation and its scenarios depend largely on the design of the FIS rules which we believe to be of great importance. In relation to that, FIS 3 rules should be formed from the perspective of an ideal supply. The drawbacks related with the use of FIS stem from errors in the definition of linguistic terms and fuzzy numbers, a process that can become complex. The role of experts is critical in capturing the characteristics of the supply chain and incorporating them into the rules.

The predictive system of active processes on metrics rates proactively supports the decision-making process by anticipating errors that may occur and provides the opportunity to perform analysis of critical or impact factors. Moreover, the prediction of future metrics rates in different circumstances approves the progress of process planning [40].

Among the main expected results are a predictive performance evaluation system focused on the sourcing area that can anticipate problems, a cyclical and adaptable measurement model for any bracket in the supply chain and the likelihood of a corrective comparison of the structure. The model identifies the current supply performance and makes a diagnosis of the structure of the measurement system used by the company with respective suggestions for improvement. It should be noted that for the application of the conceptual model, companies are free to adopt the proposed system or to continue with the usual one. The aim of the project is to provide feedback on the findings and offer guidelines or alternatives for the development of indicators.

5 Conclusions

In researching and analyzing the topic of reference in this chapter, we find how many methods, approaches, and techniques used for measurement can be adapted to the particular needs of the business. The design and selection of indicators dictate the overall performance and practices in logistics. Furthermore, it is evident that measurements can be adaptable in such a way that they can take into account qualitative, imprecise data and uncertain situations. If there are no measurements, there is no way to evaluate performance, identify errors, minimize costs, improve information flow and processes, ensure quality, etc. The theoretical review in this chapter provides an overview of supply chain performance measurement systems and can help researchers to identify and structure particular measurement systems according to the purpose of the measurement.

As relevant data, it is evident that companies place more interest in evaluations based on responsiveness, cost, sustainability, agility, the customer and internal supply chain processes, and flexibility. Multi-criteria decision techniques are highly taken into account. The most commonly used techniques are AHP, DEA, and simulation either in single or combined applications (hybrid models), and, in recent years, artificial intelligence techniques have been added to provide intelligent capabilities to systems such as the use of fuzzy logic and neural networks. On the other hand, there are the performance measurement frameworks of which the SCOR and BSC models were identified. Performance quantification is mainly based on expert judgments and information simulation.

The model and expected results corroborate the assumption that the use of predictive hybrid systems based on metrics and attributes (SCOR model with FAHP) for weight to be given and FIS for evaluations seems to be a feasible technique to help decision-makers in the performance management of the supply chain. SCOR level 1 indicators are applied as a means to assess the supply chain, allowing comparison with other chains and facilitating communication with stakeholders. The system offers the possibility to anticipate and prioritize. Additionally, the model assesses the amount of indicators that companies use and their objective with corrective effects; it is built to take into consideration the variability in the processes; it is a cyclical model to execute simulations by continuously changing the input data and the expected targets. It is easy to use, flexible, and versatile, applicable in different supply chain architectures.

In addition to that, the proposed system allows factors linking factors to improve the analysis and characterize the measurements, for which it is necessary to define fundamental aspects:

- Purpose: Prediction of performance (of lagging indicators from leading indicators).
- Strategy: The conceptual framework is designed for a generic supply chain.
- Scope of application: At the supply stage.
- Choice of metrics: Based on the SCOR model.
- Uncertainty modelling: Subjective assessment by multiple decision-makers through pairwise comparisons is proposed.
- Techniques: Hybrid approach (SCOR, FAHP, and FIS).

Three main contributions can be found in this chapter; first, it proposes a methodology related to the KPIs of the SCOR model applicable in the sourcing area. Second, by means of a benchmark between the indicators more frequently used by firms found in the literature and their relationship with the standard model, allows systems and their configurations to be analyzed and qualified. For instance, it is possible to determine whether the measurements delivered optimal results and also to evaluate the structure of the measurement systems (the number of metrics, targets, relationship with strategic objectives, etc.). Finally, the system that integrates metrics and characteristics of the SCOR model is cyclical and can be adapted to the design of the supply chain in question, a FAHP for priority analysis, and a FIS for predictive performance measurement, determining the performance characteristics with the top impact contributions to intelligent strategic decision-making and the creation of action plans.

However, the framework suffers from some limitations. Future research can address some of the following aspects: application of the model in a real case study, validation of the conceptual framework, and incorporation of neuro-fuzzy approaches in case large amounts of data are handled to train the system.

References

1. R.H. Ballou, *Business Logistics Management*,4th edn. (Prentice Hal, Hoboken, 1998)
2. I.V. Kozlenkova, G.T.M. Hult, D.J. Lund, J.A. Mena, P. Kekec, The role of marketing channels in supply chain management. J. Retail. **91**(4), 586–609 (2015)
3. Y. Qi, B. Huo, Z. Wang, H.Y.J. Yeung, The impact of operations and supply chain strategies on integration and performance. Int. J. Product. Econ. **185**, 162–174 (2017)
4. J. Jayaram, M. Dixit, J. Motwani, Supply chain management capability of small and medium sized family businesses in India: a multiple case study approach. Int. J. Product. Econ. **147**, 472–485 (2014)
5. K. Jagan Mohan Reddy, A. Neelakanteswara Rao, L. Krishnanand, A review on supply chain performance measurement systems. Proc. Manuf. **30**, 40–47 (2019)
6. A. Gunasekaran, B. Kobu, Performance measures and metrics in logistics and supply chain management: a review of recent literature (1995–2004) for research and applications. Int. J. Product. Res. **45**(12), 2819–2840 (2007)
7. P.C. Brewer, T.W. Speh, Using the balanced scorecard to measure supply chain performance. J. Bus. Logist. **21**(1), 75–93 (2000)
8. E.H. Frazelle, *Supply Chain Strategy: The Logistics of Supply Chain Management*, 1st edn. (McGraw-Hill, New York, 2002)
9. M. Christopher, *Logistics and Supply Chain Management: Strategies for Reducing Cost and Improving Service*, 2nd edn. (Financial Times/Prentice Hall, London, 1999)
10. F.R. Lima-Junior, L.C.R. Carpinetti, Quantitative models for supply chain performance evaluation: a literature review. Comput. Ind. Eng. **113**, 333–346 (2017)
11. B. Sundarakani, H.A. Razzak, S. Manikandan, Creating a competitive advantage in the global flight catering supply chain: a case study using SCOR model. Int. J. Logist. Res. Appl. **21**(5), 481–501 (2018)
12. F.R. Lima-Junior, L.C.R. Carpinetti, An adaptive network-based fuzzy inference system to supply chain performance evaluation based on SCOR® metrics. Comput. Ind. Eng. **139**, 1–19 (2020)
13. H. Balfaqih, Z.M. Nopiah, N. Saibani, M.T. Al-Nory, Review of supply chain performance measurement systems: 1998–2015. Comput. Ind. **82**, 135–150 (2016)
14. B.M. Beamon, Supply chain design and analysis: models and methods. Int. J. Product. Econ. **55**(3), 281–294 (1998)
15. A. Otto, H. Kotzab, Does supply chain management really pay? Six perspectives to measure the performance of managing a supply chain. Eur. J. Oper. Res. **144**(2), 306–320 (2003)
16. APICS - Supply Chain Operations Reference Model, version 12.0. http://www.logsuper.com/ueditor/php/upload/file/20190530/1559181653829933.pdf. Accessed 3 Mar 2017
17. S. Sipahi, M. Timor, The analytic hierarchy process and analytic network process: an overview of applications. Manage. Decis. **48**(5), 775–808 (2010)
18. J. Santos, E. Negasy, L. Cavique, Introduction to data envelopment analysis, in *Efficiency Measures in the Agricultural Sector: With Applications* (Springer, Berlin, 2013), pp. 37–50

19. E. AbuKhousa, J. Al-Jaroodi, S. Lazarova-Molnar, N. Mohamed, Simulation and modeling efforts to support decision making in healthcare supply chain management. Sci. World J. **2014**, 354246 (2014)
20. V. Belton, T. Stewart, *Multiple Criteria Decision Analysis - An Integrated Approach* (Kluwer Academic Publishers, London, 2002)
21. P. Brewer, T. Speh, Using the balanced scorecard to measure supply chain performance. J. Bus. Logist. **28**(1), 75pp. (2000)
22. O.S. Vaidya, Analytic hierarchy process: an overview of applications. Eur. J. Oper. Res. **169**(1), 1–29 (2006)
23. S. Soheilirad, K. Govindan, A. Mardani, E.K. Zavadskas, M. Nilashi, N. Zakuan, Application of data envelopment analysis models in supply chain management: a systematic review and meta-analysis. Ann. Oper. Res. **271**, 915—969 (2018)
24. G.E. Delipinar, B. Kocaoglu, Using SCOR model to gain competitive advantage: a literature review. Proc. Soc. Behav. Sci. **229**, 398–406 (2016)
25. S. Elgazzar, N. Tipi, G. Jones, Key characteristics for designing a supply chain performance measurement system. Int. J. Product. Perform. Manage. **68**, 296—318 (2019)
26. A. Najmi, M.R. Gholamian, A. Makui, Supply chain performance models: a literature review on approaches, techniques, and criteria. J. Oper. Suppl. Chain Manage. **6**, 94—113 (2013)
27. M. Keshavarz Ghorabaee, M. Amiri, E.K. Zavadskas, J. Antucheviciene, Supplier evaluation and selection in fuzzy environments: a review of MADM approaches. Econ. Res.-Ekonomska Istraživanja **30**(1), 1073–1118 (2017)
28. F. Aqlan, S.S. Lam, A fuzzy-based integrated framework for supply chain risk assessment. Int. J. Prod. Econ. **161**, 54—63 (2015)
29. L. Zanon, R. Munhoz Arantes, L. Del Rosso Calache, L. Ribeiro Carpinetti, A decision making model based on fuzzy inference to predict the impact of SCOR® indicators on customer perceived value. Int. J. Product. Econ. **223**, 1–17 (2020)
30. P. Baily, D. Farmer, D. Jessop, D. Jones, *Purchasing Principles and Management*, 9th edn. (Pearson, Boston, 2005)
31. K. Govindan, A.N. Haq, P. Sasikumar, S. Arunachalam, Analysis and selection of green suppliers using interpretative structural modelling and analytic hierarchy process. Int. J. Manage. Decis. Mak. **9**(2), 163–182 (2008)
32. P.K. Dey, W. Cheffi, Green supply chain performance measurement using the analytic hierarchy process: a comparative analysis of manufacturing organizations. Product. Plan. Control **24**(8–9), 702–720 (2013)
33. S. Luthra, D. Garg, A. Haleem, Identifying and ranking of strategies to implement green supply chain management in Indian manufacturing industry using analytical hierarchy process. J. Ind. Eng. Manage. **6**(4), 930–962 (2013)
34. J. Madaan, S. Mangla, Decision modeling approach for ecodriven flexible green supply chain, IN *Systemic Flexibility and Business Agility* (Springer, Delhi, 2015), pp. 343—364
35. S.M. Ordoobadi, Application of AHP and taguchi loss functions in supply chain. Ind. Manag. Data Syst. **110**(8), 1251—1269 (2010)
36. L. Abdullah, Fuzzy multi criteria decision making and its applications: a brief review of category. Proc. Soc. Behav. Sci. **97**, 131—136 (2013)
37. F. Farajpour, M.T. Taghavifard, A. Yousefli, M.R. Taghva, Information sharing assessment in supply chain: hierarchical fuzzy rule-based system. J. Inf. Knowl. Manage. **17**(1) (2018)
38. A. Khan, S. Kusi-Sarpong, F. Kow Arhin, H. Kusi-Sarpong, Supplier sustainability performance evaluation and selection: a framework and methodology. J. Clean. Product. **205**, 964–979 (2018)
39. U. Segundo, L. Aldámiz-Echevarría, J. López-Cuadrado, D. Buenestado, F. Andrade, T.A. Pérez, R. Barrena, E.G. Pérez-Yarza, J.M. Pikatza, Improvement of newborn screening using a fuzzy inference system. Exp. Syst. Appl. **78**, 301—318 (2017)
40. E. Domínguez, B. Pérez, Á.L. Rubio, M.A. Zapata, A taxonomy for key performance indicators management. Comput. Standards Interfaces **64**, 24–40 (2018)

41. M. Sellitto, G. Medeiros, M. Borchardt, R. Inácio & C. Viegas, A SCOR-based model for supply chain performance measurement: application in the footwear industry, International Journal of Production Research, **53**(16), 4917–4926 (2015). https://doi.org/10.1080/00207543.2015.1005251
42. P. Akkawuttiwanich, P. Yenradee, Fuzzy QFD approach for managing SCOR performance indicators. Computers and Industrial Engineering 122. 189–201 https://doi.org/10.1016/j.cie.2018.05.044

Parking Slots: The Last Mile Literature Review

Blanca Idalia Pérez-Péréz and Giovanni Lizárraga

Abstract In this study, an investigation was made of what has been applied to the problem of the so-called last mile. The delivery of the last mile is considered one of the most expensive and least efficient sections of the supply chain. Designing the last mile system efficiently is very important to serve customers efficiently and economically. They comment that the challenges that logistics and cargo transportation have faced are increasingly complicated due to transitions in the economic structure, city design, urbanization, the transportation system, and the external situations typical of the logistics activities in urban areas. Despite the fact that logistics is an important generator of employment, the negative aspects that arise in cargo transportation have increased. These situations are pollution, congestion, and inefficient use of resources. These inefficiencies can cancel your long-term benefits. Attention to the "last mile problem" leans or tends towards the best allocation of resources so that the level of service is maximized and costs are minimal in the final segment of transport. In this chapter, we talk about the parking slot problem for urban distribution and some solutions. The following document provides a literary review of logistics in the cities, their problems, and how they have tried to solve the last mile problem. The sections of this chapter are divided into introduction, problem statement, and state of the art for a local case study.

Keywords Logistics · Supply chain · VRP · Smart cities · Parking slots · Last mile

B. I. Pérez-Péréz (✉) · G. Lizárraga
Universidad Autónoma de Nuevo León, San Nicolás de los Garza, México
e-mail: blanca.perezprz@uanl.edu.mx; giovanni.lizarragalz@uanl.edu.mx

F. Torres-Guerrero et al. (eds.), *2nd EAI International Conference on Smart Technology*, EAI/Springer Innovations in Communication and Computing,
https://doi.org/10.1007/978-3-031-07670-1_10

1 Introduction

The logistics that occur in a city are a major problem for urban centers affected by the growth and consolidation of economic and industrial activities with traditional and human activities. The latest majority of studies follow the traditional transport approach with the aim of explaining variables related to transport supply instead of analyzing real demand [1].

In 2017 [2], it is commented that the challenges that logistics and cargo transportation have faced are increasingly complicated due to the transitions in the economic structure, the design of cities, urbanization, the transportation system, and the external situations proper of logistical activities in urban areas.

In urban areas, we note that the increase in freight transport vehicles and current trends in their growth are responsible for a critical increase in traffic congestion, air pollution, noise, and other externalities. This is usually replicated in the main cities of the world. New technologies and organizational strategies allow us to more efficiently manage the delivery of the last mile in urban areas [3].

Another key aspect is delivering products on time. In this aspect is where the logistics specialists have difficulties, since the merchandise must be delivered in a profitable and sustainable way for the organization. Existing literature that talks about last mile delivery manages the efficient use of resources, operations, distances, and time [4].

Despite the fact that logistics is an important generator of employment, the negative aspects that arise in cargo transportation have increased. These situations are pollution, congestion, and inefficient use of resources. These inefficiencies can cancel long-term benefits [2].

The awareness of public authorities related to urban logistics has also been increasing, and in this way, the level of public regularization can be increased. In some cases, it is through parking restrictions, limited access to certain areas, time windows, and truck restrictions in cities [5]. When authorities propose short-term regulations, such measures can increase other costs or transfer costs to other geographic areas [6].

According to [7], urban logistics is centered in three elements: (1) vehicles and flow of goods, (2) characteristics of the goods, and (3) focus of the investigation. Within the first element, vehicles and flows of goods can be analyzed independently or jointly.

The 28% of the total transportation costs of a product are attributed to the final section of the supply network [8]. Attention to the "last mile problem" leans or tends towards the best allocation of resources so that the level of service is maximized and costs are minimal in the final segment of transport [9].

Double parking is a bad practice that is commonly used by drivers to save time in search of "available bay" and unload their products to the delivery's client. However, this leads factors like road congestion, noise, insecurity, and air pollution from other vehicles blocked by the transport in question to increase. For local government zones, there are bay zones available for parking. Nonetheless, if the availability of these spaces is insufficient, drivers park in prohibited zones in an attempt to improve their productivity [10].

Due to this issue, authorities from big locations are forced to increase the number of available bays, in addition to regulating its use to free them up for as long time as possible [11].

The logistics problem of the last mile has increased in importance. The information that was found is described below.

2 Theorical Framework

Urbanization generates big populations that grow constantly, bringing forth mobility, pollution, and residue management problems. There are new information and communication technologies that unite to form solutions for smart cities [12]. Some concepts will now be mentioned.

2.1 *The Last Mile*

The delivery of the last mile is considered one of the most expensive and at the same time least efficient sections of the supply chain. Designing the last mile system efficiently is very important to serve customers efficiently and economically [13].

2.2 *Smart City*

It is a concept of utmost importance for the scientific community [14], since planning the transformation of cities of a sustainable type is a task of vital importance in the development of cities [15]. The architecture of smart cities can be structured in four layers: application, middleware, network, and detection [16]. The evolution of a low-carbon energy system is a vital part of achieving sustainable development, where transport is an extremely important link for long-term carbon reduction [17].

2.3 *VRP*

VRP, or the vehicle routing problem, is a variation of the extensively investigated traveling salesman problem. The feature of fluctuating travel duration enables VRP to account for the current conditions such as urban congestion, where the traveling speed is not constant due to variation in traffic density [18].

3 State-of-the-Art Problem Statement

3.1 Free Slots Algorithmic Forecast Design

There are many factors to consider in order to find parking spots in cities. It is subject to a low percentage of variability which depends on local regulations (subject to a slow rate of variability). (This variability is due to local regulations.) Other conditions that need to be considered are important events, day of week (normal or holiday), weather conditions, time of year, and unexpected situations like floods; the number of vehicles varies. As a consequence, there is a random distribution in the free parking spaces (the availability of parking spots). Therefore, the available parking places are considered in an algorithm as a stochastic process, where the number of places is a random variable, which is also a function of the travel time with the variables to be defined [19].

An example was applied in Zaragoza, Spain. This city has a population of 664,938 inhabitants [20]; thus, we can generalize the results to any medium-sized city. The parking data are among those selected on the platform [21]. Four-month data were used for training or analysis, while data from 6 weeks were used to verify the forecast algorithm. Verification is explained in subsection B in this section. In this way, we analyzed parking data from September 2017 to January 2018 (Fig. 1).

The following factors were considered to classify the different patterns:

- Working days versus weekends (Saturdays and Sundays) and holidays.
- Days with adverse weather (rain, wind, etc.).
- As a data mining tool for the analysis of historical data, the IBM SPSS Modeler application was used. As an example of the analysis, we present the results

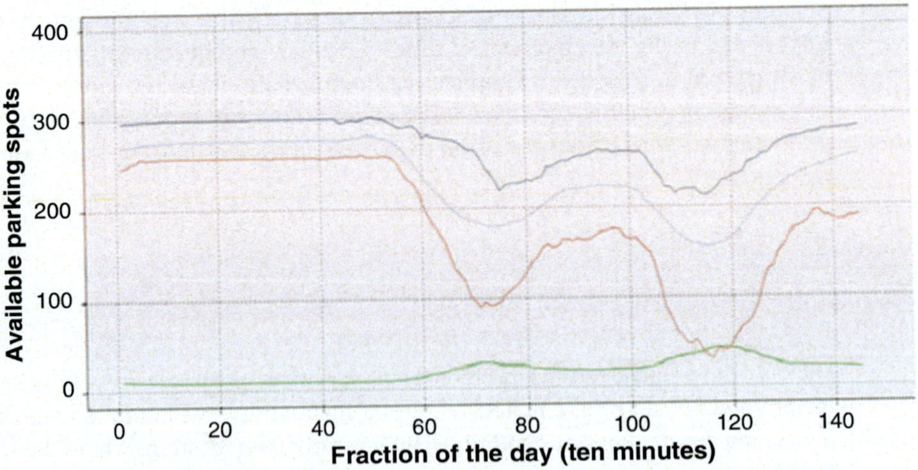

Fig. 1 Parking patterns for working days in parking lot #1 in Zaragoza

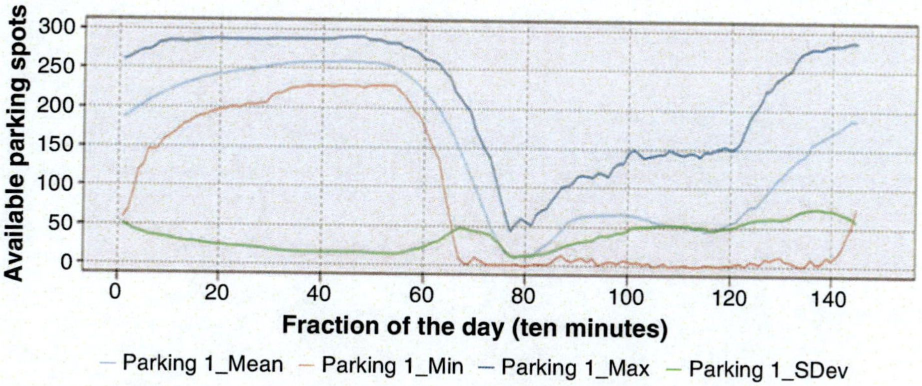

Fig. 2 Parking patterns for weekends and holidays in parking lot #1 in Zaragoza

Table 1 Forecast algorithm results

Type of day	Number of days	MPE
Working days	28	5.23%
Weekends	8	4.58%

MPE is the mean percentage error

obtained on working days and compared them with the ones obtained on weekends and holidays. The Christmas period was excluded (Fig. 2).

We considered the mean percentage error (MPE) as a measure of the accuracy of our forecast algorithm. MPE is defined as shown in Eq. (1), where at is the actual value, ft is the forecast, and n is the number of times the variable has been forecasted [22].

$$\text{MPE} = \frac{100\%}{n} \sum_{t=1}^{n} \frac{a_t - f_t}{a_t} \tag{1}$$

The results are shown in Table 1.

An optimization model for parking slot rent [23] comments that local governments stimulate sharing parking spots [24]. The proposal is a mathematical optimization method for establishing where to put reserved slots at disposal of the users. The contributions are:

- An overview of the regulations and policies of a set of major cities where carsharing has been introduced, particularly highlighting the importance of parking policies in making carsharing a success and discussing the Italian case mining.
- A mathematical optimization model to represent the decision problem of the council of a municipality that must choose which parking slots to rent to carsharing companies in a city. A linear programming problem, which includes

Fig. 3 Parking patterns for weekends and holidays in parking lot #1 in Zaragoza

Boolean variables to represent the possibility of renting or not a cluster of parking slots [24]. The following key performance indicators are managed:

- If the slot is rented or not (not rented to leave space for crashed vehicles or crashing services)
- Maximum number of rentable slots, that does not exceed the number of available slots
- Function objective

An example of the application of this model was in Rome, Italy, in 2019 (Fig. 3).

3.2 ACO: Ant Colony Optimization

It is the most common technique for merchandise delivery; however, it has the characteristic of assuming that customers will be at home or receive the product in their warehouses in a pre-established period of time, lacking flexibility. It is possible that the client is not found or cannot receive the product in his warehouse due to an emergency or the transport exceeds the pre-established time range due to traffic or some adverse situation. In this case, a second delivery would have to be scheduled which is inefficient and time consuming. To counteract the problem of inflexibility, the shared reception box delivery mode is used in cities with high population densities. This system consists of managing an intelligent reception box that is installed in a public area of the communities or at the entrance of a building or industrial park. When the products are delivered, a text message is automatically sent to the corresponding client, which contains a unique code with which the box can be opened and thus can be collected in the client's available time. Ant colony optimization, or ACO, helps build an effective and efficient delivery route. This algorithm simulates the feeding behavior of the ant colony by releasing pheromones in its pathways, which can provide heuristic information for other ants. Hormone density increases if more ants walk the same path, building the best path to the food source [25].

Attended Home Delivery (AHD). The inherent feature of AHD is that customers have to be at home during a prearranged time period and the deliveries have to be achieved within that time period. This mode is required under some circumstances; for example, the purchased goods have to be examined due to its high value or need to be signed for delivery confirmation. However, such a delivery mode lacks flexibility. Customers may not be available when the delivery arrives due to some emergency or the deliveries could not arrive due to the traffic issue. In that case, the courier may have to conduct a second delivery, which has low efficiency and is time consuming.

Shared Reception Box (SRB). When the courier delivers the purchased goods to SRB, SRB can automatically send a text message to the corresponding customer, which contains a one-off code for opening a corresponding box. After that, the corresponding customer can pick up the goods from SRB at their available time. The application of SRB can release the time constraints for both customers and couriers obviously. Moreover, the application of SRB can protect customers' privacy. Instead of using home address when purchasing online, customers can use the SRB as the consignment address. More and more e-commerce enterprises deploy SRBs in large cities so as to facilitate their business [30].

Figure 4 shows a map with the location of the 20 clients to visit (light blue diamonds). In the center is the warehouse or starting point (red square) of the delivery vehicle. The green triangles represent the savings zone. The colored dates are the seven routes that the method throws. The optimal solution purchased with AHD mode consists of seven routes, due to limitations of vehicle capacity and customer time windows [13].

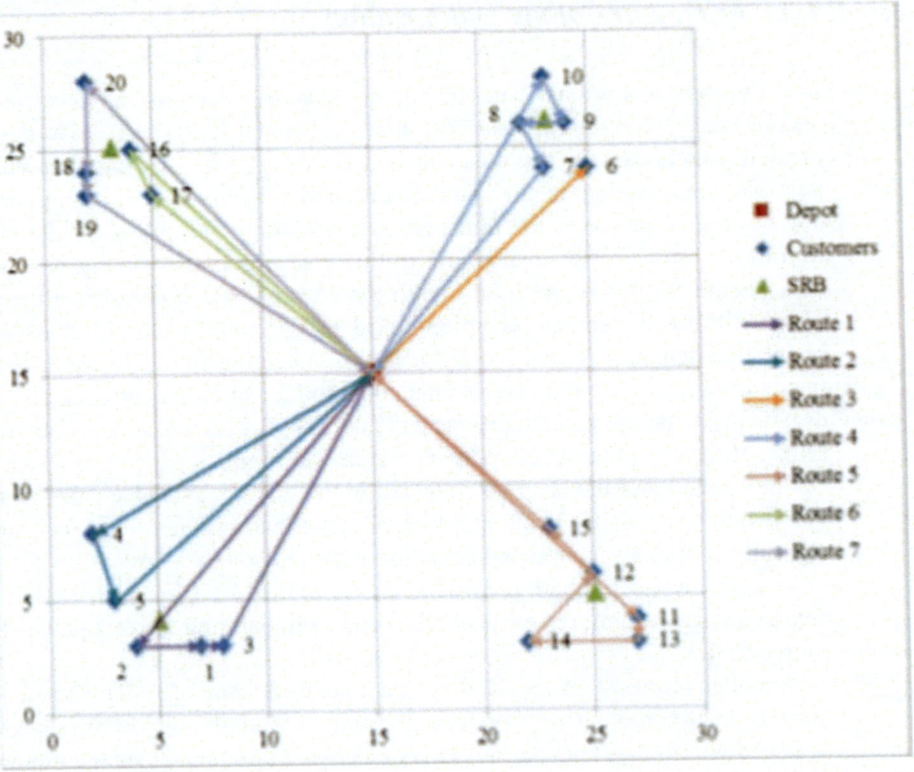

Fig. 4 Ant colony

In recent years, a boom in artificial intelligence applications and services has been observed, ranging from the management of a personal assistant to complex surveillance systems that use audio and video. With the spread of the Internet of Things, millions of devices connect to the Internet, generating millions of bytes of information on the network [13].

Another option is shared parking applications in high congestion cities, such as Beijing, Shanghai, and Hangzhou, in China. It is also possible to prepare instructions for shared parking, for example, in the Standing Committee of the Beijing People's Congress (2017) where the Regulation of Parking Management for Motor Vehicles was disclosed to the public, where it is recommended to open spaces to the public assigned that are owned by government agencies, companies, institutions, and private users with the shared pattern supported by applications, for example, Airparking, solving small-scale parking problems [26].

The literature also describes vehicle routing problem (VRP) options. A taboo algorithm is described to solve traffic and have a specific waiting time for clients. He also talks about joint trips to lower costs. A collective delivery method is managed in the last mile assisted by an application where the person closest to the customer makes the delivery. With this, the payment to the person carrying out the delivery to the end customer is the minimum possible. The mathematical model and pseudocode to define delivery routes is also presented [18].

3.3 Policies in the Logistics Sector

The implementation of policies in the logistics sector could generate unexpected side effects that sometimes undermine the performance of key economic activities of logistics operators, especially in areas such as transport service, sustainability of operations, etc. Especially for the latter, there is a lack of understanding of how sustainability performs if retail logistics solutions are affected by policy implementation and, in turn, by operator response measures.

A sustainability analysis of proven and innovative retail logistics solutions is presented that addresses the research question: "What are the effects of retail logistics solutions on overall cost and sustainability performance?" The analysis is carried out in a framework based on indicators and on the key components of sustainability (economy, environment, society), enriched by the addition of the transport component. The framework evaluates three different scenarios together with a business such as Urban Consolidation Center, Anchorage, and Shared Bus [27].

Completing urban cargo deliveries is increasingly challenging in congested urban areas, especially when delivery trucks are required to meet deadlines. Depending on the characteristics of the route, electric assistance cargo bikes can serve as an economically viable alternative to repair trucks. The purpose of this document is to compare delivery route cost offsets between box delivery trucks and electric assistance cargo bikes that have the same route and delivery characteristics and to explore the question, under what conditions do they work electric assistance cargo bikes at a lower cost than typical delivery trucks? The independent variables, constant variables, and assumptions used for the cost function comparison model are collected through data collection and a review of the literature. A delivery route was detected in Seattle and considered as a base case. The same route was modeled using electric assistance cargo bikes. Four separate delivery situations were modeled to assess how the following independent route characteristics would affect the cost of the delivery route: distance between a distribution center and a neighborhood, number of stops, distance between each stop, and number of packages per stop. The analysis shows that three of the four modeled route characteristics determined the cost offsets between repair trucks and electric assistance cargo bikes. Electric assistance cargo bikes are more cost-effective than delivery trucks for affected deliveries to the distribution center (less than 2 miles for the observed delivery route with 50 packages per stop and less than 6 miles for the hypothetical delivery route with 10 packages per stop) and in which there is a high density of residential units and low volumes of delivery per stop [28].

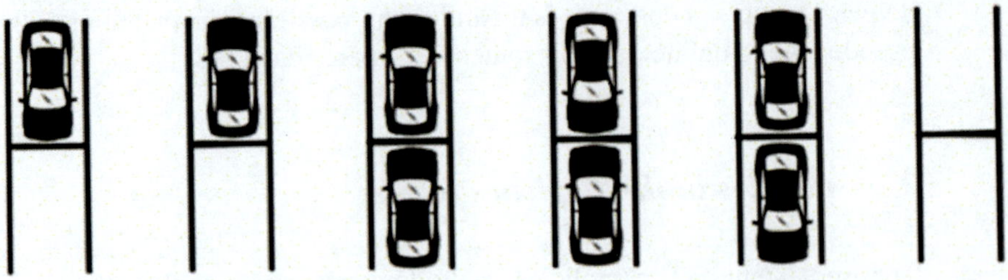

Fig. 5 Types of oriented parking lot

3.4 Forward and Reverse Parking Lot

Another aspect to consider to save time is parking methods: straight ahead or reverse. Reverse parking is considered a safer way to park in the cargo shipment, as it tries to make the environment safer when the driver leaves the parking space and can see their surroundings clearly. In a study by the National Highway Traffic Safety Administration (NHTSA), on average, 76% of drivers park straight ahead, the way many of us are used to. This situation makes sense to some extent. For example, in a double-row parallel parking lot, people have a high probability of not parking in reverse when the opposite parking space is already occupied; in a single-row parking lot, people are less willing to reverse parking because it would take more time and reduce speed of the vehicles behind them as seen in Fig. 5 [29].

3.5 Predicting Bay Available

In this part, a smart system that is capable of recognizing its surroundings, learn, and make decisions. It can predict the vacancy of parking slots in real time from the origin to the delivery area and vice versa. First, it predicts available slots in real time and notifies the user. The prediction methods used were the following: linear regression, gradient boosting, random forest, and neural networks. These methods analyzed the history of occupied spaces to predict their future availability. The experiment took place with historical data from 2017 from the city of Melbourne where 4300 ground sensors were installed in 303 segments of 35 areas of the central business district. The different techniques gave the following graphics of Fig. 6 [10].

When analyzing the error metrics, the method with the best results (lowest MSE) was random forest with 1.69%, followed by regression tree with 2.10%. In the third place was the artificial neural network that in addition had a high computational time of 6 hours against the 12 minutes maximum that other methods took, getting 2.62%. In the last place was stacking with 5.41%. For most of these models, predictions are more exact when available spaces are fewer or equal to 50. The performance of this

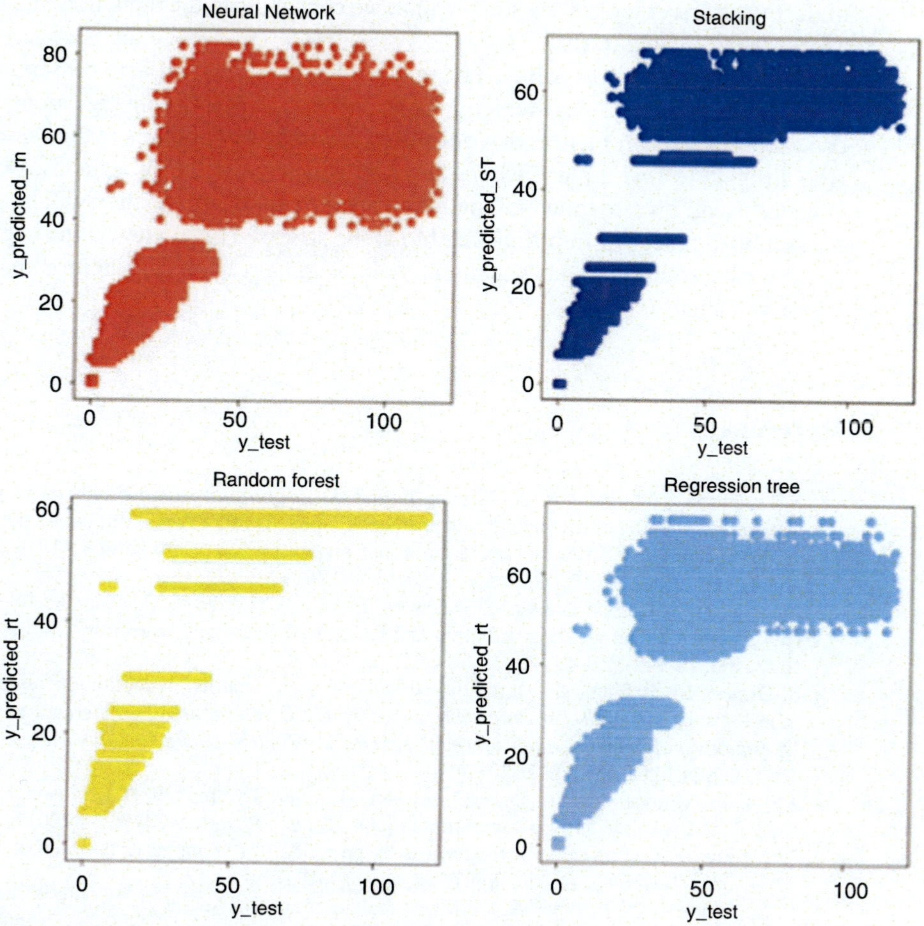

Fig. 6 Comparative graph between the predictions obtained by some techniques and the real value

case is conditioned to the drivers' respect of their reservation and times of arrival. With this, if required, a penalty to drivers that do not comply with their schedule is suggested so as to not risk the system's function [10].

4 Conclusions

Freight parking is a serious problem in big cities. As cities grow, the merchandise transportation problem increases with them. So, finding effective methodologies that help us increase the number of available slots for merchandise delivery in the last mile is of utmost importance, and just like that, we could contribute to the reduction of the time a driver looks for a slot. With it, we contribute to the reduction of gasoline expenses, noise, and traffic.

Currently, we want to solve a last mile problem in a central area of San Nicolás de los Garza, Nuevo León, Mexico, where various businesses converge along with a central plaza. The problem is that deliveries are made by suppliers at the same time, who struggle to find a place where they can park their delivery vehicles. Delivery vehicles usually waste time looking for a parking slot, or worse: they will be parked in prohibited places or in places that block the streets. This review is the beginning to put forth a solution to this situation. We hope to have the results approximately in a year and a half.

References

1. J. Sepúlveda, J.W. Escobar, W. Adarme-Jaimes, Un algoritmo para el problema de ruteo de vehículos con entregas divididas y ventanas de tiempo (SDVRPTW) aplicado a las actividades de distribución de PYMEs del comercio al por menor. DYNA (Colombia) **81**(187), 223–231 (2014). https://doi.org/10.15446/dyna.v81n187.46104
2. I. Cardenas, Y. Borbon-Galvez, T. Verlinden, E. Van de Voorde, T. Vanelslander, W. Dewulf, City logistics, urban goods distribution and last mile delivery and collection. Compet. Regul. Netw. Ind. **18**(1–2), 22–43 (2017)
3. S. Digiesi, M. P. Fanti, G. Mummolo, B. Silvestri, Externalities reduction strategies in last mile logistics: a review, in *Proceedings – 2017 IEEE International Conference on Service Operations and Logistics, and Informatics, SOLI 2017*, *2017-Janua*, 248–253. https://doi.org/10.1109/SOLI.2017.8121002 (2017)
4. J. Holguín-Veras, I. Sánchez-Díaz, C. Lawson, M. Jaller, S. Campbell, H. Levinson, H.S. Shin, Transferability of freight trip generation models. Transp. Res. Rec. **2379**, 1–8 (2013)
5. J.L. Routhier, C. Raux, An attractiveness-based model for shopping trips in urban areas, in *12th World Conference on Transport Research*. (2010a)
6. E. Verlinde, N. De Laender, S. De Maesschalck, et al., The social gradient in doctor-patient communication. Int. J. Equity Health **11**, 12 (2012). https://doi.org/10.1186/1475-9276-11-12
7. J. González-Feliu, M.G. Cedillo-Campos, J.L. García-Alcaraz, An emission model as an alternative to OD matrix in urban goods (2014)
8. R.W. Goodman, Whatever you call it, just don' t think of, in *Global Logistics and Supply Chain Strategies*, *December*, 1–6 (2005)
9. B. Balcik, B.M. Beamon, K. Smilowitz, Last mile distribution in humanitarian relief. J. Intell. Transp. Syst. Technol Plan. Oper. **12**(2), 51–63 (2008). https://doi.org/10.1080/15472450802023329
10. H. Errousso, N. Malhene, S. Benhadou, H. Medromi, Predicting car park availability for a better delivery bay management. Procedia Comput. Sci. **170**, 203–210 (2020). https://doi.org/10.1016/j.procs.2020.03.026
11. A. Comi, B. Buttarazzi, M. Schiraldi, Smart urban freight transport: tools for planning and optimising delivery operations, in *Simulation Modelling Practice and Theory*, vol. 88, (Elsevier Ltd, 2018), pp. 48–61. https://doi.org/10.1016/j.simpat.2018.08.006
12. D. Bruneo, S. Distefano, M. Giacobbe, A. Longo Minnolo, F. Longo, G. Merlino, D. Mulfari, A. Panarello, G. Patanè, A. Puliafito, C. Puliafito, N. Tapas, An IoT service ecosystem for Smart Cities: the #SmartME project. IoT **5**, 12–33 (2019). https://doi.org/10.1016/j.iot.2018.11.004
13. L. Zhou, Y. Lin, X. Wang, F. Zhou, Model and algorithm for bilevel multisized terminal location-routing problem for the last mile delivery. Int. Trans. Oper. Res. **26**(1), 131–156 (2019). https://doi.org/10.1111/itor.12399

14. R.K.R. Kummitha, N. Crutzen, How do we understand smart cities? An evolutionary perspective. Cities **67**, 43–52 (2017)
15. S.C. Addanki, H. Venkataraman, Greening the economy: a review of urban sustainability measures for developing new cities. Sustain. Cities Soc. **31**, 1–8 (2017)
16. T. Watari, B.C. McLellan, D. Giurco, E. Dominish, E. Yamasue, K. Nansai, Total material requirement for the global energy transition to 2050: a focus on transport and electricity. Resour. Conserv. Recycl. **148**, 91–103 (2019)
17. A. Alkhamisi, M.S.H. Nazmudeen, S.M. Buhari, A cross-layer framework for sensor data aggregation for IoT applications in smart cities, in *Proceedings of the 2016 IEEE International Smart Cities Conference (ISC2), Trento, Italy*, 12–15 September 2016
18. H. Desai, B. Remer, B. Chalaki, L. Zhao, A. Malikopoulos, J. Rios-Torres, *Vehicle Routing for the Last-Mile Logistics Problem*. http://arxiv.org/abs/1904.05464 (2019)
19. J. Barbancho, J. Ropero, J. Luque, A. Caraballo, C. León, Social parking: applying the citizens as sensors paradigm to parking guidance and information. Sustainability (Switzerland) **11**(23), 1–19 (2019). https://doi.org/10.3390/su11236549
20. Ayuntamiento de Zaragoza. Movilidad. 2018. Available online: http://www.zaragoza.es/ciudad/viapublica/movilidad/estacionamientos/. Accessed 12 July 2018
21. El Tiempo en Zaragoza. Histórico. 2018. Available online: https://www.eltiempo.es/zaragoza.html?v= historico. Accessed 10 December 2018
22. S. Makridakis, S.C. Wheelwright, R.J. Hyndman, *Forecasting: methods and applications*, 3rd edn. (Wiley, New York, NY, USA, 1998)
23. R. Dowling, J. Kent, Practice and public-private partnerships in sustainable transport governance: the case of car sharing in Sydney, Australia. Transp. Policy **40**(3), 58–64 (2015)
24. S. Carrese, F. D'Andreagiovanni, T. Giacchetti, A. Nardin, L. Zamberlan, An optimization model for renting public parking slots to carsharing services. Transp. Res. Procedia **45**, 499–506 (2020). https://doi.org/10.1016/j.trpro.2020.03.064
25. Zhang, Y. Guo, W., Zhang Xu, M., Li, L Z. (2016). Parking spaces repurchase strategy design via simulation optimization. J. Intell. Transp. Syst. 20 (3), 255–269
26. H. Xiao, M. Xu, How to restrain participants opt out in shared parking market? A fair recurrent double auction approach. Transp. Res. C Emerg. Technol. **93**(June), 36–61 (2018). https://doi.org/10.1016/j.trc.2018.05.023
27. K. Papoutsis, W. Dewulf, T. Vanelslander, E. Nathanail, Sustainability assessment of retail logistics solutions using external costs analysis: a case-study for the city of Antwerp. Eur. Transp. Res. Rev. **10**(2) (2018). https://doi.org/10.1186/s12544-018-0297-5
28. M. Sheth, P. Butrina, A. Goodchild, E. McCormack, Measuring delivery route cost trade-offs between electric-assist cargo bicycles and delivery trucks in dense urban areas. Eur. Transp. Res. Rev. **11**(1) (2019). https://doi.org/10.1186/s12544-019-0349-5
29. G. Mangano, G. Zenezini, The value proposition of innovative last-mile delivery services from the perspective of local retailers. IFAC-PapersOnLine **52**(13), 2590–2595 (2019). https://doi.org/10.1016/j.ifacol.2019.11.597
30. S.Z. Zhang, C.K.M. Lee, *Flexible vehicle scheduling for Urban Last Mile Logistics: The emerging technology of Shared Reception Box. 2016 IEEE International Conference on Industrial Engineering and Engineering Management (IEEM)*.https://doi.org/10.1109/ieem.2016.7798211 (2016)

Technologies in Education for Visually Impaired People: A Literature Review

Néstor Ulises López Flores and Aída Lucina González Lara

Abstract The advancement of technology allows better assistive tools to be developed for people with visual impairment. In this chapter, a literary review is presented on the technologies that are used to support the education of people with visual impairments and describe about the seriousness and the impact of inclusive education which indicates equality among students regardless of the challenges they may have; the main technologies used are screen readers which allow the screen of the device to be heard through audio and the screen magnifiers which enlarge the text to make it easier to be read by the user with visual impairment; both technologies allow access to the mass media which help their education; finally, it is intended to obtain the main characteristics that these technologies must have for the future development of a new educational tool for visually impaired people.

Keywords Visual impairment · Inclusive education · Accessibility · Assistive technologies

1 Introduction

The International Classification of Functioning, Disability and Health (ICF) defines disability as all impairments, activity limitations, and participation restrictions. Disability is the interaction between individuals with a health condition and environmental factors [1].

Visual impairment is defined as a condition that directly involves the functioning of the eye, and it comes from moderate difficulty in perceiving lights to total blindness; therefore, the scope of information perceived by a person from the environment is limited, so this significantly compromises the physical integrity of those who suffer [2].

N. U. López Flores (✉) · A. L. González Lara
Universidad Autónoma de Nuevo León, San Nicolás de los Garza, Mexico
e-mail: nestor.lopezfl@uanl.edu.mx; aida.gonzalezlr@uanl.edu.mx

F. Torres-Guerrero et al. (eds.), *2nd EAI International Conference on Smart Technology*, EAI/Springer Innovations in Communication and Computing,
https://doi.org/10.1007/978-3-031-07670-1_11

According to the World Health Organization in 2010, there were almost 40 million people with visual disabilities in the world, which 80% of them are in working age; unfortunately, people with visual disabilities face challenges to reach the labor market, for example, some schools are not sufficiently adapted, so people with visual impairment cannot finish their studies; also, this population has problems with inaccessible public transport and urban mobility [3].

The inclusion of people with disabilities in the use of technologies can have a great impact for them since technologies can be developed or adapted according to the needs and characteristics of each user. In this case, users with visual impairments will benefit from technologies because they can provide information in another sense and would not necessarily depend on visual elements [4].

These adaptive technologies and tools can have a great impact and increase the orientation, mobility, and independence of visually impaired people and can help by providing information about the environment through other sensory means [5].

Inclusive education refers to vulnerable groups, that is, giving equality and providing the same educational services as non-vulnerable students; therefore, it implies that educational institutions require infrastructure, Internet, tools, and different types of support as well as that teachers have adequate equipment and training to teach quality teaching in order to achieve greater inclusion [6].

In this work, we will be searching for technologies and tools developed for people with visual impairment for their inclusion in education; later, we will describe the most important characteristics of these technologies for the future development of more accessible technologies for people with visual impairment.

2 Methodology

An international bibliographic review of the articles published in the databases was carried out through the National Consortium of Scientific and Technological Information Resources (CONRICYT), applying a time limit of 5 years (2015–2020). We also searched in Google Scholar applying the same time limit.

The keywords used were "discapacidad," "visual," "educación," "tecnologías," "inclusión," and "accessibilidad" in Spanish and "disability," "visual," "accessibility," "technologies," "inclusive," and "education" in English. The following compositions were made: "visual impairment," "inclusive education," "accessible education," "teaching method," and "visual impairment technology."

The methodology of the bibliographic search was based on the collection of documents that talk about developed technologies to support teaching methods for people with visual impairments.

We selected the research papers whose focus were aimed to aid people with total blindness.

3 Results

The screen reader is an adaptive technology which is currently the main resource that allows people to access information and perform reading and writing tasks. This tool helps a student with visual impairment to have autonomy with reading a book, searching for information, downloading music, printing documents, writing letters, correcting errors with the spoken program, and having email, among other activities. Most common screen readers such as Non-Visual Desktop Access (NVDA) and Job Access With Speech (JAWS) are frequently used by interactive learning systems for visually impaired people [7].

Barros, Carrión, Cedillo, Idrovo, López, and Maldonado developed a Java application where the main objective is to teach students with visual impairments; the application is an audible multiple option questionary, and the user uses the numpad from the keyboard to answer the question; they used the numpad because the key "5" has a braille ubication to help the students [8].

Othmani, González, Rodrigo, and Perez made a study to value the distance education for 60 students with visual impairment, and they founded out positive results in their questionaries and interviews, but something to achieve this goal is that there should be full accessible interface to take class and accessible materials [9]. Juárez, Aquino, and Garcia in 2015 found that only five universities from Mexico offer distance educative programs to people with visual impairments, so this means that Mexican educative institutions should focus more to this type of programs to achieve a better inclusion [6].

Matoušek et al. developed a web-based system to facilitate access to educational materials by reading. Also, the system enables teachers to prepare and process arbitrary topics focusing on technical documents that contain mathematics and physics formulas. This was made for lower secondary school in the Czech Republic. The system converts the content automatically to speech, and it was evaluated by 41 students with visual impairments and 3 teachers. They had positive results, especially for the difficult topics [10].

Vera, Marcillo, and Pereira developed a system that helps blind people to learn their environment and be able to navigate in indoors or outdoors scenarios. The prototype is portable so they can use it when needed. It was evaluated by five people with visual impairment. The system can identify obstacles in real-life scenario, which is very helpful, and it is a low-cost solution for visually impaired people. Their study also found that it can be used in the rehabilitation of people who recently became blind. The prototype they developed showed positive results with aiding people with visual impairments to navigate [11].

Khan et al. proposed a mail system called TetraMail, an accessible blind-friendly email client; this mail system organizes the content of the screen in manageable partitions of five sections and rearranges the activities of an email in these five sections; it was evaluated by 38 blind people by performing 14 email activities; they compared this email client with other existent email clients, and the results demonstrate that TetraMail have a better user interaction experience [12].

Ferrand, Alouges, and Aussal developed an embedded device capable of guiding people using spatialized virtual sound source; with this device, they have been able to guide people to do sports like running or roller skating in a protected environment; using sound stimuli, their experiments showed positive results with blind users [13].

Cardillo, Li, and Caddemi developed a system that employs a radar that can be attached to a traditional white cane to aid navigation of visually impaired people. It detects obstacles in the scanned area, and it also discerns between human and nonhuman obstacles by detecting their breathing vital sign. This could have a great impact for visually impaired people specifically in the ways these people learn to navigate in new areas [14].

Iswahyudi, Anam, and Sujanarko developed a visual aid tool for visually impaired people based on a convolutional neural network; the system is focused on object detection and object positioning; it accepts video that is connected to a camera; voice commands are spoken by the user when he searches for an object; then the speech input is recognized by the device, and the system guides the user to find the object needed through the audio speaker output; this tool will help the visually impaired people to identify objects and the positions of them; this will help their independence [15].

Alkhalid, Kadhim Oleiwi, and Muhsin proposed a system based on face detection to aid visually impaired people to navigate independently; the system used Haar cascade algorithm and OpenCV using python; their results showed great effectiveness and efficiency; they will be working on detecting objects and vehicles too; this will cause a great impact for the independence of visually impaired people [16].

Aisy and Eliyawati proposed "Bluino," a walking support for visually impaired students; it is equipped with an ultrasonic HC-SR04 sensor to detect an object within 50 cm, which triggers a sound; this tool is attached to a box and placed on the user's leg; Bluino can help the visually impaired to detect objects and help their navigation [17].

Ersanty, Wibisono, Niratama, and Sasongko made a comparison of two common screen readers (JAWS and NVDA) and found that it does not influence on learning process outcome; these tools are only to support visual translation to audio forms; they made this comparison with Surabaya State University visually impaired students [18].

Lutfun Nahar, Riza Sulaiman, and Azizah Jaafar developed a low-cost system to learn math braille using Nemeth Braille and calculating numbers called "NC tutor" which provides voice and vibrational feedbacks to assist the users; it was developed mainly in JAVA, and it was evaluated by teachers, experts, and students and provided good results in these students in Bangladesh; it also had great results in usability tests [19].

Vetter proposed WELLE, a web-based music environment for blind people; this tool is developed in JavaScript and runs on standard browsers; this web application is text based and centered around an input field, which serves as the main interaction element; all the elements are accessible to a screen reader; it was tested in a 3 h workshop with a group of six pupils aged 10–14 years, whom five were blind and one was visually impaired; also, two blind teachers participated in the workshop and showed positive results. WELLE is a work in progress and not yet a stable music environment but offers blind people quick and uncomplicated access to musical drafts and playful engagements with sounds through a textual interface [20].

Malaver and Molina developed a learning system that is made by virtual objects using Exelearning, this virtual objects are formed by an accessible video; in this tool, it was focused in teaching biology; in this particular case, the diversity of birds in Cundinamarca; in their results, it showed that students with visual impairments felt comfortable using the system, it also showed that most of the users think that this system helps to enhance their knowledge [21].

Cecily Morrison, Nicolas Villar, Anja Thieme, Zahra Ashktorab, Eloise Taysom, Oscar Salandin, Daniel Cletheroe, Greg Saul, Alan F Blackwell, Darren Edge, Martin Grayson, and Haiyan Zhang presented Torino, a tangible programming language for teaching programming concepts to children regardless of the level of vision; a tangible programming language uses physical objects to represent or interact with programming constructs; this makes a great way to teach programming concepts to visually impaired people; these tools were designed using an inclusive design; this means that it is not only for a specific type of users; it can be easily used by users with or without disabilities [22].

When reviewing the previous studies, we can now mention the main characteristics that a technology should have to help people with visual disabilities in their education; these characteristics are the following:

- The tool must be fully accessible with a screen reader.
- The tool should be accessible in any kind of device.
- You should be able to use the tool without having to be in a specific place.
- The tool should have a good usability.
- A multisensorial tool is highly recommended.

4 Conclusions

Present literature reviews show that multiple technologies to aid visually impaired people have been developed; most of these technologies are divided in two sections, that is, orientation and mobility and educational; it becomes a focus in the community to achieve inclusive education and an independence for visually impaired people.

In the Mexico 2020 population census, it was reported that there are more than 20 million people who have a disability; that is more than 16% of the entire population of the country. In this census, it was also reported that more than 12 million people have a visual problem, making this the most common disability in Mexico; therefore, it is necessary to develop new assistive technologies to aid people who are visually impaired.

The discovery of new technologies in teaching methods not only helps in the inclusive education, but they also help to increase their confidence. The development of these teaching methods will help students with visual impairment to focus on the topics they prefer and not only the topics that are naturally accessible to them.

Multisensorial technologies not only help people with disabilities, but they also help to see and learn in different ways, so these technologies can enhance the knowledge of people without disabilities; in this literature review, the visually impaired people benefit more from the technologies that use audio instead of visual elements.

References

1. World Health Organization, Disability and health [Online] (2018). Available: https://www.who.int/news-room/fact-sheets/detail/disability-and-health. Last Accessed 29 Nov 2020
2. J.Q. Beltrán Ramírez, J. Zepeda Gómez, M. Maciel Arellano, V. Larios Rosillo, J. Espinoza, J. Martínez Mendoza, Tecnologías en apoyo al traslado y acceso a la información destinado a personas con discapacidad visual, I, vol. 14, n.° 26, pp. 70–78, abr. 2019
3. J.T. Kaiser, J.L. Cmar, S. Rosen, D. Anderson, *Scope of practice in orientation and mobility. Association for Education and Rehabilitation of the Blind and Visually Impaired O&M Division IX* (Association for Education and Rehabilitation of the Blind and Visually Impaired, Alexandria, VA, 2018)
4. A.L. Esparza-Maldonado, L.Y. Margain-Fuentes, F.J. ÁlvarezRodríguez, E.I. Benítez-Guerrero, Desarrollo y evaluación de un Sistema Interactivo para personas con discapacidad visual. TecnoLógicas **21**(41), 149–157 (2018)
5. A. Ferreira Gomes Filho, R. De Toledo, Visual Management and Blind Software Developers, 2015 Agile Conference, Washington, DC, 2015, pp. 31–39. https://doi.org/10.1109/Agile.2015.14
6. D. Juárez, S. Aquino, V. García, Educación a distancia para alumnos con discapacidad visual: estado actual en el ámbito de la educación superior en México. Integración. Revista sobre Discapacidad Visual No.67. **67**, 1887–3383 (2015)
7. P.d.J. Cortés García, L.D.S. Barrera, M.C.C. García, La importancia de las Tecnologías de la Información y Comunicación (TIC), en la Inclusión Educativa de las Personas con Discapacidad en la República Mexicana: The importance of Information and Communication Technologies (TIC) in the Educational Inclusion of Persons with Disabilities in the Mexican Republic. EDU **15**(16), 127–139 (2017)
8. J. Barros, L. Carrión, Á. Cedillo, J. Idrovo, A. López, S. Maldonado, Aplicación informática para asistir a niños y jóvenes con discapacidad visual (2018)
9. O. Othmani, M.L. Cacheiro-González, C. Rodrigo-San Juan, V.A. Lorenzo Pérez, Accesibilidad del modelo de educación a distancia para estudiantes con discapacidad visual. Revista de Educación Inclusiva **11**(1), 25–38 (2018)
10. J. Matoušek, Z. Krňoul, M. Campr, et al., Speech and web-based technology to enhance education for pupils with visual impairment. J. Multimodal User Interfaces **14**, 219–230 (2020). https://doi.org/10.1007/s12193-020-00323-1
11. D. Vera, D. Marcillo, A. Pereira, Blind guide: anytime, anywhere solution for guiding blind people, in *Recent Advances in Information Systems and Technologies. WorldCIST 2017*, Advances in Intelligent Systems and Computing, ed. by Á. Rocha, A. Correia, H. Adeli, L. Reis, S. Costanzo, vol. 570, (Springer, Cham, 2017). https://doi.org/10.1007/978-3-319-56538-5_36
12. A. Khan, S. Khusro, B. Niazi, et al., TetraMail: a usable email client for blind people. Univ. Access Inf. Soc. **19**, 113–132 (2020). https://doi.org/10.1007/s10209-018-0633-5

13. S. Ferrand, F. Alouges, M. Aussal, An electronic travel aid device to help blind people playing sport. IEEE Instrum. Meas. Mag. **23**(4), 14–21 (2020)
14. E. Cardillo, C. Li, A. Caddemi, Empowering blind people mobility: a millimeter-wave radar cane. In *2020 IEEE International Workshop on Metrology for Industry 4.0 & IoT*, pp 213–217 (IEEE, 2020)
15. Iswahyudi, K. Anam, B. Sujanarko, Development of a visual aid tool for blind people based on faster R-CNN. In *AIP Conference Proceedings* vol 2278, No. 1, 020049 (AIP Publishing LLC, 2020)
16. FF Alkhalid, BK Oleiwi, MA Muhsin "Real Time Blind People Assistive System Based on OpenCV", JUBES, vol. 28, no. 2, pp. 25 - 33, Nov. 2020.
17. R. Aisy, E.C. Prima, E. Eliyawati, Bluino: Blind Arduino Sensor to Assist Students with Visual Impairment
18. D. Ersanty, S.S. Wibisono, F. Niratama, T.B. Sasongko, Comparison of JAWS and NVDA as assistive technology for college students with special needs at Universitas Negeri Surabaya. JPI (Jurnal Pendidikan Inklusi) **3**(2), 136–146 (2020)
19. L. Nahar, R. Sulaiman, A. Jaafar, An interactive math braille learning application to assist blind students in Bangladesh. Assist. Technol. (2020). https://doi.org/10.1080/10400435.2020.1734112
20. J. Vetter, WELLE-a web-based music environment for the blind. NIME'20, July 21–25, 2020, Royal Birmingham Conservatoire, Birmingham City University, Birmingham, United Kingdom
21. A.D. Malaver Baéz, P.A. Molina Miranda, Diseño e implementación de un Objeto Virtual de Aprendizaje para la enseñanza inclusiva de la diversidad de las aves presentes en el departamento de Cundinamarca, Colombia enfocado a personas con discapacidad visual vinculadas al Instituto Nacional para Ciegos INCI
22. C. Morrison, N. Villar, A. Thieme, Z. Ashktorab, E. Taysom, O. Salandin, D. Cletheroe, G. Saul, A.F. Blackwell, D. Edge, M. Grayson, H. Zhang, Torino: a tangible programming language inclusive of children with visual disabilities. Human–Comput. Interact. **35**, 1 (2018). https://doi.org/10.1080/07370024.2018.1512413

sEMG Classification of Upper Limb Movements Under Different Loads

Arturo González-Mendoza, Alberto-Isaac Perez-Sanpablo, Ivett Quiñones-Urióstegui, R. López-Gutíerrez, and Sergio Salazar-Cruz

Abstract Surface Electromyography (sEMG) might provide new ways of communication or control with devices or surroundings through the use of a Human-Machine Interface (HMI). Recent researches have determined that a critical challenge in sEMG-based HMIs is how often a classification algorithm classifies a body movement through sEMG correctly in real-time. This article aims to determine if it is possible to identify motor gestures through sEMG signals generated by different loads, to investigate the features that facilitate the classification of sEMG signals. Four machine learning classifier models based on Decision Tree (DT), Support Vector Machine (SVM), K-Nearest Neighbor (KNN), and ensemble algorithms were evaluated. The input data to feed the proposed algorithms is in the time and frequency-domain of sEMG signals generated by different loads. Furthermore, this paper identifies the features of sEMG that facilitate the classification of sEMG signals generated by different loads through a Principal Component Analysis (PCA). Results on classifier models based on KNN and SVM show to have the best accuracy performance. Results show that time-domain features are reliable for gesture recognition based on sEMG signals. The results also show that gesture recognition thought sEMG signals generated by different external loads is possible. This might be important since if muscle activation can be detected. A gesture recognition system can be applied, helping people with a disability.

A. González-Mendoza (✉)
National Institute of Rehabilitation Luis Guillermo Ibarra Ibarra, Mexico City, Mexico

National Council for Science and Technology, CONACYT-CINVESTAV, Mexico, Mexico

A.-I. Perez-Sanpablo · I. Quiñones-Urióstegui
National Institute of Rehabilitation Luis Guillermo Ibarra Ibarra, Mexico City, Mexico
https://www.inr.gob.mx/i17.html

R. López-Gutíerrez · S. Salazar-Cruz
National Council for Science and Technology, CONACYT-CINVESTAV, Mexico, Mexico
e-mail: jesusl.lopez@cinvestav.mx

F. Torres-Guerrero et al. (eds.), *2nd EAI International Conference on Smart Technology*, EAI/Springer Innovations in Communication and Computing,
https://doi.org/10.1007/978-3-031-07670-1_12

Keywords Surface electromyography · Machine learning · Principal component analysis

1 Introduction

Surface electromyography (sEMG) has been used as input for Human–Machine Interfaces (HMI) to recognize the user's body movements and to translate them into machine commands. Some applications of sEMG-based HMI in upper limb rehabilitation include bionic hands [12, 37], rehabilitation devices [1], and assistive devices [30]. One of the most critical challenges in sEMG-based HMIs is how often the algorithm classifies a body movement through sEMG correctly (accuracy) in real-time. Accurate real-time response is needed to give people adequate afferent feedback [5]. According to [10], no differences in accuracy are observed when time windows of analysis between 32 ms and 256 ms are used. The classification process is better known as pattern recognition, which has three fundamental stages: signal pre-processing, feature extraction, and classifier model training [40]. The second and third stages of pattern recognition are known as gesture recognition. Signal pre-processing includes hardware and software denoising, full-wave rectification, and smoothing to get the envelope of the sEMG signal [24]. Feature extraction plays an essential role in classification accuracy and feasibility. Several features from time-domain, frequency-domain, and time-frequency-domain have been used to analyze sEMG signals [16]. Due to the huge amount of available features, sometimes methods such as PCA have been used to reduce feature space dimensionality [25] [28]. On occasions, computationally simpler classification methods are preferred over more accurate but more complex ones due to their low processing time, making them more feasible to be applied in real-time [3, 29]. For instance, time-domain features which are less computational complex to calculate are preferred over frequency-domain and time-frequency-domain features [8, 21]. Training of numerous classifier models has been performed on features extracted from sEMG like Gaussian mixture models [7], support vector machines SVM [39], k-nearest neighbors KNN [29], DTs [14], Random Forest (RF) or ensemble methods [4]. Classifier models have reported similar results (up to 98% accuracy) as long as classifier models are adjusted correctly and use a good set of features [18]. The use of classifier models has allowed recognition of the direction of motion of single-Degrees of Freedom (DoF) and multi-DoF movements of the upper limb. For example, direction of eight single-DoF and multi-DoF movements of elbow, forearm, and wrist (elbow flexion/extension, forearm pronation/supination, wrist radial/ulnar deviation, and wrist flexion/extension) were identified using nine time-domain features (Root Mean Square (RMS), Mean Absolute Value (MAV), Wave Length (WL), number of zero crossings (ZC), number of slope sign changes (SSC), and four autoregressive coefficients (AR)) recorded from eight muscles (biceps brachii, triceps brachii, pronator teres, supinator, Flexor Carpi Radialis (FCR), Extensor Carpi Ulnaris (ECU), Extensor Carpi Radialis (ECR), and ECU) using Linear Discriminant Analysis (LDA) [17] . After reducing the feature set, McDonald et al. concluded that two features of the sEMG (RMS and MAV) are enough for detecting movement direction with a 100% accuracy in nine able-bodied subjects. Sun et al. [32] compared four machine learning classifier models (KNN, Artificial Neural Network (ANN), RF, and SVM) to identify four wrist movements (wrist extension/flexion, making a fist, and resting) using forty-two time-domain, frequency-domain, and time-frequency-domain features calculated from sEMG signals of four wrist muscles (Flexor Carpi Ulnaris (FCU), ECR, ECU, and abductor pollicis longus (APL)). After optimizing channel and feature selection, Sun [32] showed that muscle selection has an impact on classifier model accuracy. An accuracy above 96.77% was found for seven, time-domain and time-frequency-domain features on one muscle in nine able-bodied subjects (ECU), using the RF algorithm. Also, ensemble methods that combine multiple classifier models to increase accuracy have been used [9, 41]. In physical therapy, resistance exercises, where the user feels the effect of additional loads while performing movement, are frequently used [34]. According to [36] the influence of external loads on muscle motor control is reflected on sEMG signals. Many sEMG-based HMIs have been designed with good results, but none has considered the effect of different loads felt by the user [6, 15, 33]. This work aims to analyze the accuracy of classification algorithms to identify loading conditions based on sEMG features. As a result, this work's main contributions are comparing and identifying four classifiers (DT with kernel medium Tree, KNN model with kernel functions medium KNN, SVM model with kernel functions medium Gaussian, and the ensemble model Boosted Trees) and a set of features to identify loading conditions based entirely on sEMG data accurately. Time-domain and frequency-domain sEMG features generated under the influence of different external loads are compared to identify best sEMG features that facilitate the classification of sEMG.

2 Methods

The proposed methodology aims to acquire and identify single-DoF upper limb movements and external loading conditions based on sEMG features. With the purpose of making a clear description of the methodology section is divided into the following subsections: Measurement Setup describes the mechanical configuration of equipment used to control movement mechanics for the acquisition of sEMG signals. Data acquisition, where the protocol for the acquisition of the signals, is described. Signal pre-processing, a section where signal processing is described. Pattern Recognition, a section that describes the metrics extracted from the sEMG signals. Features reduction test, where a series of tests are proposed to identify the contribution of the features.

2.1 Measurement Setup

In order to acquire the sEMG signals generated by single-DoF upper limb movements under different external loads condition, three measurement setups were proposed (See Fig. 1). The measurement setups are based on Von Werder's methodology [36]. In all the measurement setups, a pulley machine (Lojer, Finland) is used to apply a constant external load to the joint understudy along with its full range of motion. A deflection pulley with a radius of 4 cm is connected to the pulley machine. In the same axis of the deflection pulley, an elbow flexion/extension, wrist flexion/extension, or wrist medial/lateral deviation joint pulleys can be connected. The joint pulleys allow aligning the joint to measure to be aligned with the deflection pulley. In the measurements setups of the elbow flexion/extension, and wrist flexion/extension the subject under study sits on a chair in an upright position in parallel to the pulley machine and is asked to align the joint to measure with the specific joint deflection pulley (See Fig. 1a, b). In the case of the wrist medial/lateral deviation measurement setup, the subject under study is asked to sit in front of the pulley machine, and then the subject is asked to align the joint to the wrist medial/lateral deflection pulley (see Fig. 1c). In the cases of the wrist flexion/extension, and medial/lateral deviation measurement setups, the user was asked to rest the arm on a table, and then the arm was fixed with Velcro straps to the table to avoid compensatory movements. To avoid joint misalignment, the pulley machine was adjusted in height to the user. Von Werder et al. [36], the sEMG studied elbow flexion and extension movements by acquiring sEMG signals of muscles biceps brachii, brachioradialis and triceps muscles. Since in this paper additionally wrist flexion/extension, and wrist medial/lateral deviation are studied, the muscles Triceps Brachii Lateral Head (TBLAH),FCR, FCU, ECR, ECU, were chosen due to their relevance for the designated movements.

2.2 Data Acquisition

The sEMG signals were acquired at a sampling rate of 1000 Hz, with bipolar Ag-AgCl circular electrodes with a 2 cm diameter, using the commercial sEMG system Datalog (Biometrics, Newport ,UK). The distance between electrodes was 2 cm. The Biceps Brachii Long Head (BBLH), and Triceps Brachii Long Head (TBLH) electrode placement, was done according to Surface ElectroMyoGraphy for the Non-Invasive Assessment of Muscles (SENIAM) [31]. TBLAH, FCR, FCU, ECR, ECU, Brachioradialis (BRA), electrodes were placed according to [19], because there are no SENIAM recommendations for their placement. The kinematics of the elbow and wrist movements were recorded at a sampling frequency of 100 Hz with twelve flex13 cameras Optitrack System (Natural Point, Corvallis Oregon, USA). Reflective markers were placed according to the standard Baseline Upper Body (25) biomechanical marker set [22].

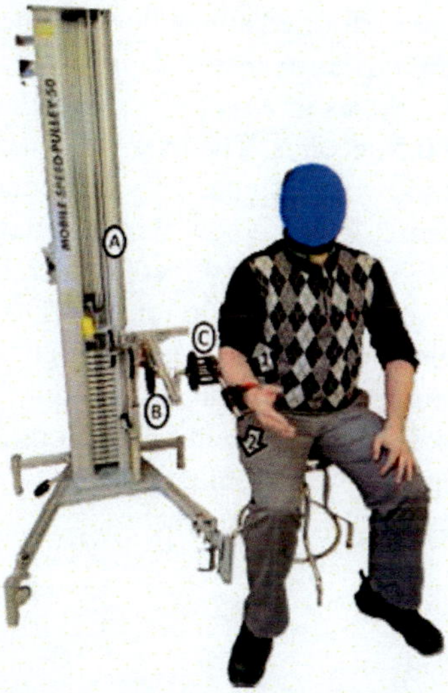

(a) Elbow (1)flexion/(2)extension measurement setup.

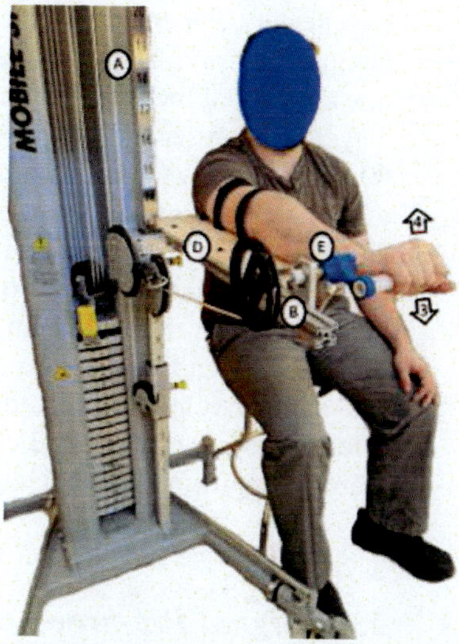

(b) Wrist (3)Flexion/(4)Extension measurement setup

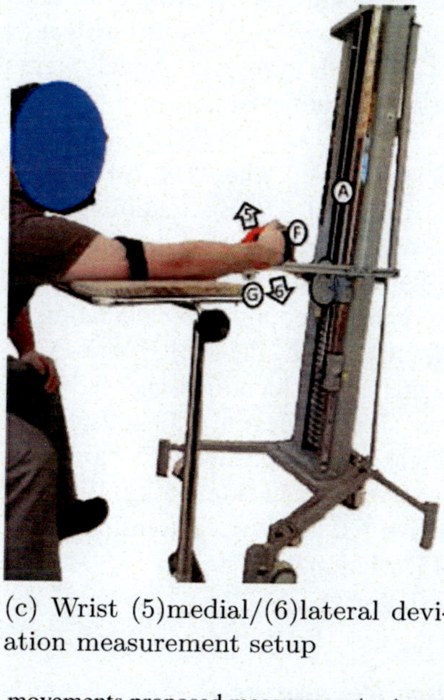

(c) Wrist (5)medial/(6)lateral deviation measurement setup

Fig. 1 single-DoF upper limb movements proposed measurement setups. (**a**) Pulley machine. (**b**) Pulley machine—deflection pulley. (**c**) Elbow flexion/extension deflection pulley. (**d**) Arm fixation table for wrist flexion/extension measurement setup. (**e**) wrist flexion/extension deflection pulley. (**f**) Wrist medial/lateral deflection pulley. (**g**) Arm fixation wrist medial/lateral deviation table

Data acquisition was performed on a subject with an age of 29 years old considered healthy. For each movement, the user performed elbow flexion/extension, wrist flexion/extension, and wrist medial/lateral deviation over the entire range of motion of each joint at the following cadences: 3 cycles in 2 seconds, 4 cycles in 1 second, 5 cycles in 0.5 seconds and, 6 cycles in 0.5 seconds. The user was asked to rest for 60 seconds between each set of cycles to avoid fatigue. This procedure was done with the following movements and external loads: elbow flexion/extension at 0.5, 2.5, 5 kg, wrist flexion/extension at 0.5, 2.5 kg, and wrist medial/lateral deviation at 0 and 0.5 kg. External loads of 0.5, 2.5, and 5 kg equal to torques of 0.08, 0.4, and 2 Nm, respectively, for a 4 cm radius deflection pulley. Loads were chosen based on the subject's capabilities to perform a full set of maneuvers without fatigue [36]. The angular velocities related to the cycles of movement range from 20 to 280°/s. These speeds were selected considering a slow movement performed by a subject who suffers from a movement pathology to a healthy subject.

2.3 Signal Pre-processing

The joint angles were obtained with an inverse kinematic analysis to the marker-set trajectories obtained with the Optitrack system. The inverse kinematic analysis was performed with the OpenSim [27] software using a biomechanical model previously presented in [11]. Then joint angle signals were resampled to a frequency of 1000 Hz. Finally, a second-order Butterworth low pass filter with a $f_{-3db} = 5$ Hz was applied.

The sEMG signals were filtered with a bandpass fourth-order digital Butterworth filter with a $f_{-3db} = 40$–450 Hz. Later on, the sEMG were full-wave rectified and normalized to one with respect to the maximum value. A low pass second-order Butterworth filter with $f_{-3db} = 10$ Hz was applied. The resampled angle and sEMG signals were synchronized through calculation of correlation between joint's angle and sEMG raw data from the BBLH and FCR. The minima and maxima of the joint angle signals were identified and used to split and label the data. The used data labels were: Elbow Flexion 0.5 kg, Elbow Extension 0.5 kg, Elbow Flexion 2.5 kg, Elbow Extension 2.5 kg, Elbow Flexion 5.0 kg, Elbow Extension 5.0 kg, Wrist Flexion 0.5 kg, Wrist Extension 0.5 kg, Wrist Flexion 2.5 kg, Wrist Extension 2.5 kg, Wrist Flexion 5.0 kg, Wrist Extension 5.0 kg, Wrist Medial Deviation 0.0 kg, Wrist Lateral Deviation 0.0 Kg.

2.4 Gesture Recognition

Once the data had been segmented and labeled, a mobile window of 256 ms was created, and features from time-domain and frequency-domain were calculated. The used features were selected for their frequent use in gesture recognition [17, 20,

32, 38, 40]. The selected features were the Integrated EMG (IEMG) [20], the MAV [13, 20, 32, 40], the Simple Square Integral (SSI) [20], the Variance (VAR) [20, 32], the RMS [13, 20, 32, 40], the WL [13, 20, 32, 40, 40], the Median Frequency (MDF) [20, 32], the Mean Frequency (MNF) [20, 32]. The selected features are defined next:

$$IEMG = \sum_{n=1}^{N} |X_n| \tag{1}$$

$$MAV = \frac{1}{N}\sum_{n=1}^{N} |X_n| \tag{2}$$

$$SSI = \sum_{n=1}^{N} |X_n|^2 \tag{3}$$

$$VAR = \frac{1}{N-1}\sum_{n=1}^{N} {X_n}^2 \tag{4}$$

$$RMS = \sqrt{\frac{1}{N}\sum_{n=1}^{N} {X_n}^2} \tag{5}$$

$$WL = \sum_{n=1}^{N-1} |X_{n+1} - X_n| \tag{6}$$

$$\sum j = 1MDFP_j = \frac{1}{2}\sum j = 1MP_j \tag{7}$$

$$MNF = \frac{\sum_{j=1} Mf_j P_j}{\sum j = 1MP_j} \tag{8}$$

where X_n is the sEMG signal in a segment. N is the total length of the sEMG signal. P_j is the power spectrum at a frequency $binj$. Finally f_j is the frequency of spectrum at frequency $binj$. n and j are the $n - sample$ and $j - bin$, respectively. The proposed features were evaluated in four classification models: DT with kernel medium Tree, KNN model with kernel functions medium KNN, SVM model with kernel functions medium Gaussian, and the ensemble model Boosted Trees. The total number of evaluated samples was of 320,649 samples. Due to the large size of the dataset, the 50% held out validation method was used. The feature extraction was done in MATLAB (MathWorks Inc, Massachusetts, USA). The classification models were trained with the classification learner toolbox from MATLAB. The classifiers were selected based on previous works explained in the introduction section. The measure used to compare the different classifier models was accuracy. Accuracy (Acc) [23] is defined in Eq. 9, where TP are True Positives and, TN are True Negatives.

$$Acc = \frac{\sum TP + \sum TN}{\sum TP + \sum TN + \sum FP + \sum FN} \tag{9}$$

2.5 *Features Reduction Test*

The appropriate selection features from processed sEMG data have the purpose of obtaining a good class separability. A good class separability ensures a low misclassification rate. This stage of pattern recognition is identified as an essential part since the best discrimination of features leads to the best performance in a system recognition process [40]. Seven experimental methods were proposed in order to identify those features that best achieve class separability of sEMG. The proposed tests are described next:

- The first test evaluated time-domain features, without considering the frequency-domain features. In this test, the time-domain features were calculated for the eight muscle signals. A total of forty-eight features were evaluated, and fourteen classes were taken into account.
- The second test consisted of using as a feature only the frequency-domain features. Sixteen features were evaluated, and fourteen classes were taken into account.
- The third test considered the time-domain features, and the frequency-domain features were added. A total of sixty-four features were evaluated, and fourteen classes were taken into account.
- The fourth test was carried out after a feature reduction analysis with a PCA. This test consisted of obtaining the results of the classifier models with the feature of IEMG. A total of eight features were evaluated, and fourteen classes were taken into account.
- The fifth test uses the reduced set of six classification classes, where the external loads are not taken into account. In this test, only time-domain features are tested. This test is done to compare results with previous studies that do not take an external load into account.

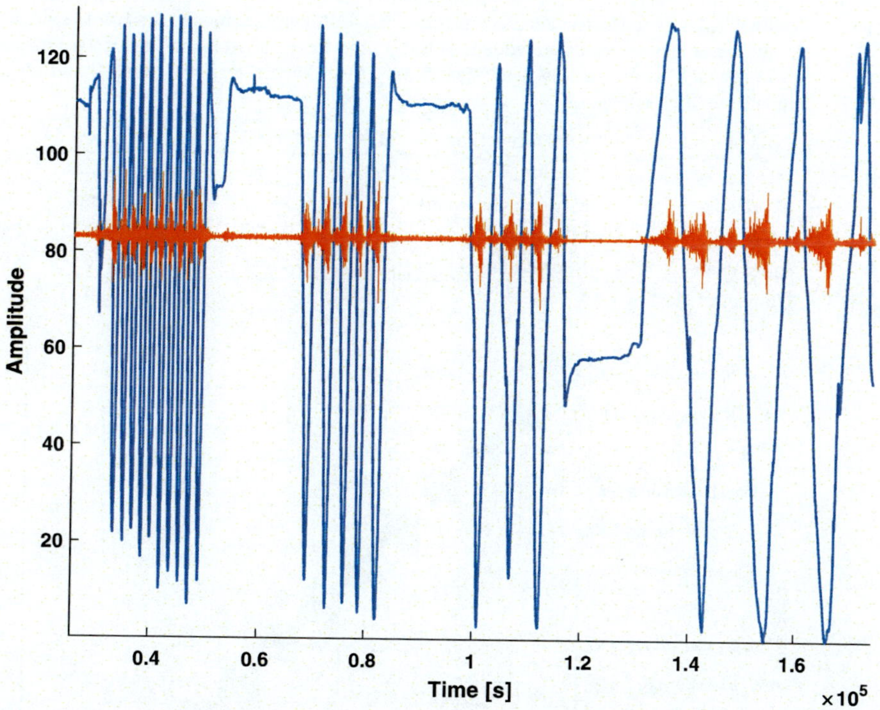

Fig. 2 [Elbow flexion/extension and sEMG 5.0 Kg external load, elbow joint range of movement (blue color) and sEMG signal (red color) synchronization. The sEMG signal was amplified with a gain of 40, and an offset of 80 was added, with the purpose to facilitate the visual comparison of the synchronization process

3 Results

As it was mentioned in the methodology section, the resampled joint angle and sEMG signals were synchronized. The result of the synchronization process is shown in Fig. 2. The average number of samples per class is 22904 with and standard deviation of 4006.9 samples. The results of all the evaluated classifier models and test are shown on Table 1. A disadvantage of accuracy is that if data is unbalanced, it could yield misleading results. Observing the behavior of classifier models through confusion matrices is good practice, i.e., in Figs. 3 and 4 shows how the percentage of true positives increases based on accuracy. The results of the PCA are normalized to one and are shown in Fig. 5.

Table 1 Classification test accuracy results. Test 1 only uses time-domain features. In test 2 only frequency domain features are added. In test 3 time-frequency-domain features are added. In test 4 only IEMG feature is only used. Test 5 only time-domain features are used and no loads are considered in classification

Type of model	Kernel function	Test 1	Test 2	Test 3	Test 4	Test 5
DT	Medium Tree	62.3	45.7	69.2	52.0	79.1
SVM	Medium Gaussian	95.9	44.0	96.8	90.5	94.4
KNN	Medium KNN	99.7	71.3	99.8	99.5	95.7
Ensemble	Boosted Trees	66.2	47.6	83.9	56.9	84.7
Number of classes		14	14	14	14	6

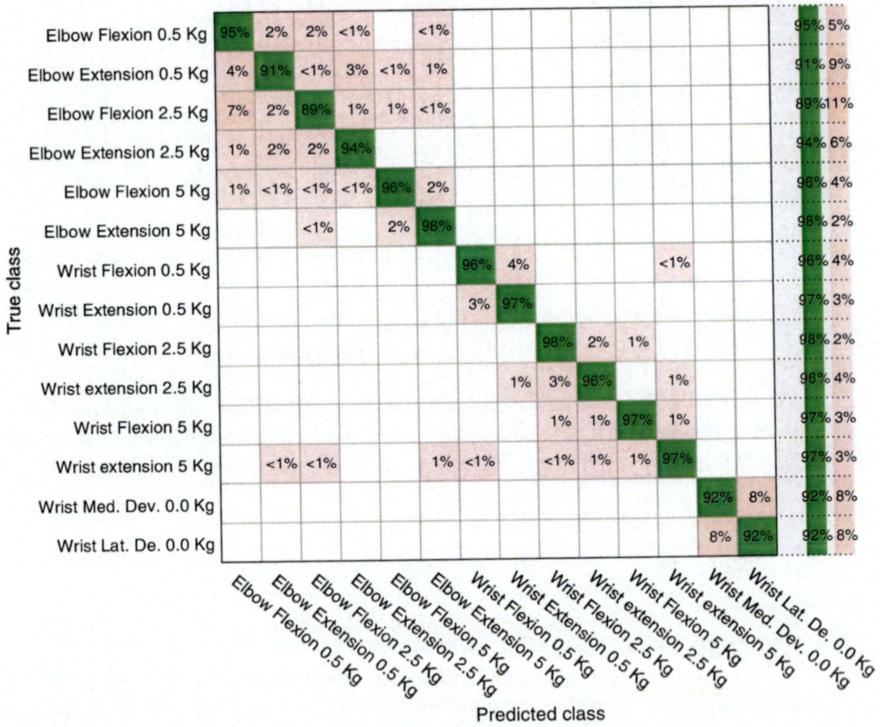

Fig. 3 DT classifier model with medium kernel function. Confusion matrix of test 5. Accuracy of 79.1%

4 Discussion and Conclusions

This paper proposed five tests that aimed to discern for the sEMG feature contribution and investigate the extent to which a set of features can be reduced to a point where high accuracy is reached resulting in short processing time for HMIs. KNN and SVM accuracy results from test six compared with the results

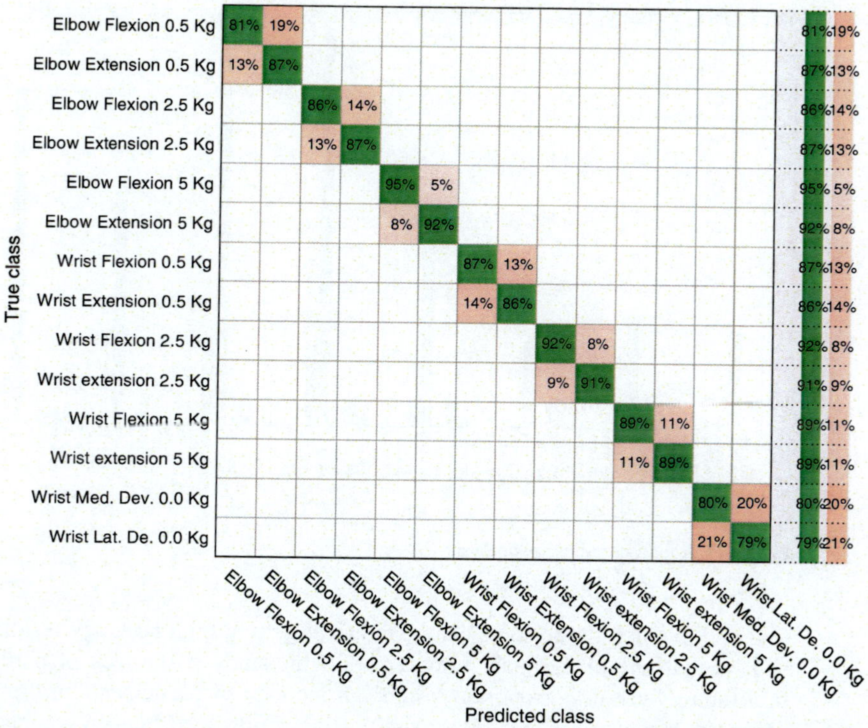

Fig. 4 KNN classifier model with medium kernel function. Confusion matrix of test 1. Accuracy of 99.7%

of Sun et al. [32], in wrist flexion/extension score similar results (over 90%), these results confirm that the classifier behaves correctly. Accuracy results of the different tests performed (see Table 1.) show that time-domain features make the main contribution, this is also supported by PCA analysis. From the performed tests and PCA analysis, is concluded that the frequency-domain features do not contribute to gesture recognition. The results explained preceding are consistent with reported results on [20, 32]. Also, comparing the positive and negative confusion matrices from test 5 of the DT, and test 1 SVM (see Figs. 3 and 4). It can be seen that the models scoring less than 90% have a high number of false positives, even in some cases, high misclassification rates are achieved. This supports the criteria of [17], where a classifier model on gesture recognition should have an accuracy of over 90%. From PCA analysis Fig. 5 can be inferred that the FCR, FCU, ECR, and ECU sEMG signals have the best performance to signal classification; this might be because more movements related to the wrist are classified and is also supported by the results reported on [32]. Comparing accuracy's like [26], showed that SVM, KNN, classifier models scored the highest scores; therefore, their use it is recommended, this is supported by [2, 25, 35]. It is also concluded from test

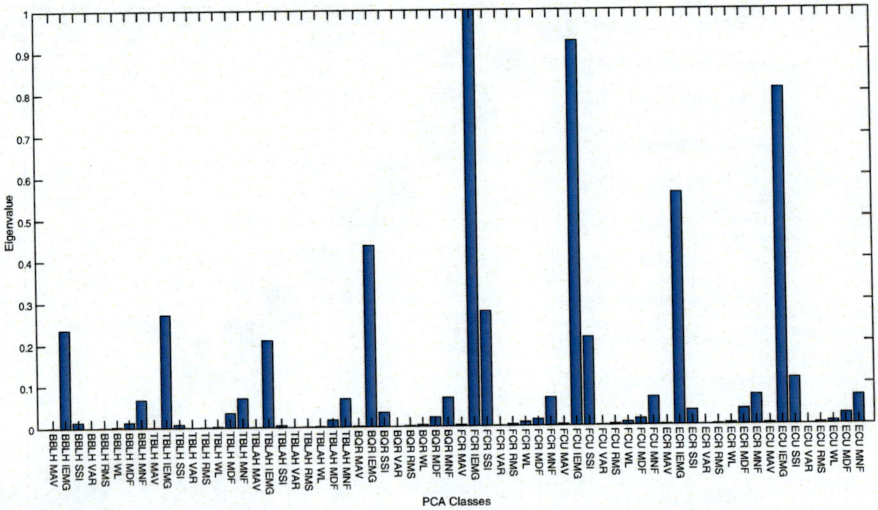

Fig. 5 PCA feature reduction analysis

four and five that the features can be reduced, getting high accuracy results. It can be said that one of the main limitations of this study is the lack of participants. In a future work to corroborate results, an increase of participants in the study is proposed, however, since similar classification results are obtained compared with other articles, as explained previously if an increase of participants is used, similar products are expected to be obtained to the result reported in this paper. From the obtained results, it can be concluded that gesture recognition thought sEMG signals generated by different external loads is possible. Low external loads can be identified; this is good results since it could help subjects with a low carrying capacity level. It can be concluded that HMI based on sEMG can be used for rehabilitation purposes.

References

1. Q. Ai, Q. Liu, W. Meng, S.Q. Xie, Chapter 2 - state-of-the-art, in *Advanced Rehabilitative Technology*, ed. by Q. Ai, Q. Liu, W. Meng, S.Q. Xie (Academic Press, Cambridge, 2018), pp. 11–32. https://doi.org/https://doi.org/10.1016/B978-0-12-814597-5.00002-3
2. M.Z. Al-Faiz, A.A. Ali, A.H. Miry, A k-nearest neighbor based algorithm for human arm movements recognition using EMG signals, in *EPC-IQ01 2010 - 2010 1st International Conference on Energy, Power and Control* (2010), pp. 159–167
3. M.S. Alam, A.S. Arefin, Real-time classification of multi-channel forearm EMG to recognize hand movements using effective feature combination and LDA classifier. Bangladesh J. Med. Phys. **10**(1), 25–39 (2018). https://doi.org/10.3329/bjmp.v10i1.39148
4. M. Atzori, A. Gijsberts, C. Castellini, B. Caputo, A.G.M. Hager, S. Elsig, G. Giatsidis, F. Bassetto, H. Müller, Electromyography data for non-invasive naturally-controlled robotic hand prostheses. Sci. Data **1**, 140053 (2014). https://doi.org/10.1038/sdata.2014.53

5. D. Blana, A.J. Van Den Bogert, W.M. Murray, A. Ganguly, A. Krasoulis, K. Nazarpour, E.K. Chadwick, Model-based control of individual finger movements for prosthetic hand function. IEEE Trans. Neural Syst. Rehabil. Eng. **28**(3), 612–620 (2020). https://doi.org/10.1109/TNSRE.2020.2967901
6. A. Blanco, J.M. Catalán, J.A. Díez, J.V. García, E. Lobato, N. García-Aracil, Electromyography assessment of the assistance provided by an upper-limb exoskeleton in maintenance tasks. Sensors **19**(15), 3391 (2019). https://doi.org/10.3390/s19153391
7. J.U. Chu, Y.J. Lee, Conjugate-prior-penalized learning of Gaussian mixture models for multifunction myoelectric hand control. IEEE Trans. Neural Syst. Rehabil. Eng. **17**(3), 287—297 (2009). https://doi.org/10.1109/tnsre.2009.2015177
8. W.M.B.W. Daud, A.B. Yahya, C.S. Horng, M.F. Sulaima, R. Sudirman, Features extraction of electromyography signals in time domain on biceps brachii muscle. Int. J. Model. Optim. **3**(6), 515–519 (2013). https://doi.org/10.7763/ijmo.2013.v3.332
9. T.G. Dietterich, An experimental comparison of three methods for constructing ensembles of decision trees: bagging, boosting, and randomization. Mach. Learn. **40**(2), 139–157 (2000). https://doi.org/10.1023/A:1007607513941
10. K. Englehart, B. Hudgins, A robust, real-time control scheme for multifunction myoelectric control. IEEE Trans. Biomed. Eng. **50**(7), 848–854 (2003). https://doi.org/10.1109/TBME.2003.813539
11. A. González-Mendoza, R. Lopéz-Gutierrez, A.I. Pérez-SanPablo, S. Salazar-Cruz, I. Quiñones-Uriostegui, M.H. Ba Tho, T. Dao, Upper limb musculoskeletal modeling for human-exoskeleton interaction, in *2019 16th International Conference on Electrical Engineering, Computing Science and Automatic Control (CCE)* (2019), pp. 1–5. https://doi.org/10.1109/ICEEE.2019.8884537
12. A.V. Ivanov, T. Skripnik, Human-machine interface with motion capture system for prosthetic control, in *2019 IEEE Conference of Russian Young Researchers in Electrical and Electronic Engineering (EIConRus)* (2019), pp. 235–239. https://doi.org/10.1109/EIConRus.2019.8657282
13. P. Kaczmarek, T. Mańkowski, J. Tomczyński, *putEMG-A surface electromyography hand gesture recognition dataset*. Sensors **19**(16), 3548 (2019). https://doi.org/10.3390/s19163548
14. M. Kurzynski, A. Zolnierek, A. Wolczowski, Control of bio-prosthetic hand via sequential recognition of EMG signals using rough sets theory. Adv. Intell. Soft Comput. **57**, 455–462 (2009). https://doi.org/10.1007/978-3-540-93905-4_54
15. H. Liu, J. Tao, P. Lyu, F. Tian, Human-robot cooperative control based on sEMG for the upper limb exoskeleton robot. Robot. Auton. Syst. **125**, 103350 (2020). https://doi.org/https://doi.org/10.1016/j.robot.2019.103350
16. F.A. Mahdavi, S.A. Ahmad, Surface electromyography feature extraction based on wavelet transform. Int. J. Integrated Eng. **4**(3), 1–7 (2013). http://penerbit.uthm.edu.my/ojs/index.php/ijie/article/view/615
17. C.G. McDonald, J.L. Sullivan, T.A. Dennis, M.K. O'Malley, A myoelectric control interface for upper-limb robotic rehabilitation following spinal cord injury. IEEE Trans. Neural Syst. Rehabil. Eng. **28**(4), 978–987 (2020). https://doi.org/10.1109/TNSRE.2020.2979743
18. G.L.E.J. Mizrahi, *Electromyography Pattern-Recognition-Based Control of Powered Multifunctional Upper-Limb Prostheses*, chap. 6 (IntechOpen, Rijeka, 2011). https://doi.org/10.5772/22876
19. A. Perotto, E.F. Delagi, *Anatomical Guide for the Electromyographer: The Limbs and Trunk* (Charles C Thomas, Springfield, 2005). https://books.google.com.mx/books?id=uwos8W4HiQ8C
20. A. Phinyomark, S. Hirunviriya, C. Limsakul, P. Phukpattaranont, Evaluation of EMG feature extraction for hand movement recognition based on Euclidean distance and standard deviation, in *ECTI-CON2010: The 2010 ECTI International Conference on Electrical Engineering/Electronics, Computer, Telecommunications and Information Technology* (2010), pp. 856–860

21. A. Phinyomark, P. Phukpattaranont, C. Limsakul, Feature reduction and selection for EMG signal classification. Exp. Syst. Appl. **39**, 7420–7431 (2012)
22. N. Point, Baseline upper body (25) (2016). https://v20.wiki.optitrack.com/index.php?title=Baseline_Upper_Body_(25)
23. D.M.W. Powers, Ailab: Evaluation: from precision, recall and f-measure to ROC, informedness, markedness & correlation. J. Mach. Learn. Technol. **2**, 37–63 (2007). http://www.bioinfo.in/contents.php?id=51
24. W. Rose, Standards for reporting EMG data. J. Electromyograph. Kinesiol. **38**, I–II (2018). https://doi.org/10.1016/s1050-6411(18)30035-x
25. S. Said, I. Boulkaibet, M. Sheikh, A.S. Karar, S. Alkork, A. Nait-ali, Machine-learning-based muscle control of a 3D-printed bionic arm. Sensors **20**(11), 3144 (2020). https://doi.org/10.3390/s20113144
26. B. Schabron, Z. Alashqar, N. Fuhrman, K. Jibbe, J. Desai, Artificial neural network to detect human hand gestures for a robotic arm control, in *2019 41st Annual International Conference of the IEEE Engineering in Medicine and Biology Society (EMBC)* (2019), pp. 1662–1665. https://doi.org/10.1109/EMBC.2019.8857264
27. A. Seth, J.L. Hicks, T.K. Uchida, A. Habib, C.L. Dembia, J.J. Dunne, C.F. Ong, M.S. DeMers, A. Rajagopal, M. Millard, S.R. Hamner, E.M. Arnold, J.R. Yong, S.K. Lakshmikanth, M.A. Sherman, J.P. Ku, S.L. Delp, OpenSim: simulating musculoskeletal dynamics and neuromuscular control to study human and animal movement. PLOS Comput. Biol. **14**(7), 1–20 (2018). https://doi.org/10.1371/journal.pcbi.1006223
28. L. Shaw, S. Bagha, Online EMG signal analysis for diagnosis of neuromuscular diseases by using PCA and PNN. Int. J. Eng. Sci. **4**(10), 4453–4459 (2012). http://doaj.org/doaj?func=fulltext&aId=1175416
29. H. She, J. Zhu, Y. Tian, Y. Wang, H. Yokoi, Q. Huang, SEMG feature extraction based on Stockwell transform improves hand movement recognition accuracy. Sensors **19**(20), 4457 (2019). https://doi.org/10.3390/s19204457
30. D.A. Sinyukov, K.L. Troy, M.P. Bowers, T. Padir, 13 - wheelchairs and other mobility assistance, in *Biomechatronics*, ed. by M.B. Popovic (Academic Press, Cambridge, 2019), pp. 373–417. https://doi.org/https://doi.org/10.1016/B978-0-12-812939-5.00013-6
31. D. Stegeman, H. Hermens, Standards for surface electromyography: the European project surface EMG for non-invasive assessment of muscles (SENIAM) **1** (2007), pp. 108–112. https://citeseerx.ist.psu.edu/viewdoc/summary?doi=10.1.1.623.2040
32. H. Sun, X. Zhang, Y. Zhao, Y. Zhang, X. Zhong, Z. Fan, A novel feature optimization for wearable human-computer interfaces using surface electromyography sensors. Sensors **18**(3), 869 (2018). https://doi.org/10.3390/s18030869
33. Z. Tang, K. Zhang, S. Sun, Z. Gao, L. Zhang, Z. Yang, An upper-limb power-assist exoskeleton using proportional myoelectric control. Sensors **14**, 6677–6694 (2014). https://doi.org/10.3390/s140406677
34. N.F. Taylor, K.J. Dodd, D.L. Damiano, Progressive resistance exercise in physical therapy: a summary of systematic reviews. Phys. Ther. **85**(11), 1208–1223 (2005). https://doi.org/10.1093/ptj/85.11.1208
35. D.C. Toledo-Pérez, J. Rodríguez-Reséndiz, R.A. Gómez-Loenzo, J.C. Jauregui-Correa, Support vector machine-based EMG signal classification techniques: a review. Appl. Sci. **9**(20), 4402 (2019). https://doi.org/10.3390/app9204402
36. S.C.F.A. von Werder, C. Disselhorst-Klug, The role of biceps brachii and brachioradialis for the control of elbow flexion and extension movements J. Electromyograph. Kinesiol. **28**, 67–75 (2016). https://doi.org/10.1016/j.jelekin.2016.03.004
37. Y. Wu, D. Jiang, X. Liu, R. Bayford, A. Demosthenous, A human–machine interface using electrical impedance tomography for hand prosthesis control. IEEE Trans. Biomed. Circuits Syst. **12**(6), 1322–1333 (2018). https://doi.org/10.1109/TBCAS.2018.2878395
38. G. Yang, J. Deng, G. Pang, H. Zhang, J. Li, B. Deng, Z. Pang, J. Xu, M. Jiang, P. Liljeberg, H. Xie, H. Yang, An IoT-enabled stroke rehabilitation system based on smart wearable armband and machine learning. IEEE J. Transl. Eng. Health Med. **6**, 2100510 (2018). https://pubmed.ncbi.nlm.nih.gov/29805919. https://doi.org/10.1109/JTEHM.2018.2822681. ISSN: 2168-2372

39. M. Yoshikawa, M. Mikawa, K. Tanaka, A myoelectric interface for robotic hand control using support vector machine, in *2007 IEEE/RSJ International Conference on Intelligent Robots and Systems* (2007), pp. 2723–2728. https://doi.org/10.1109/IROS.2007.4399301
40. Z. Zhang, K. Yang, J. Qian, L. Zhang, Real-time surface EMG pattern recognition for hand gestures based on an artificial neural network. Sensors **19**(14), 3170 (2019). https://doi.org/10.3390/s19143170
41. Z.H. Zhou, *Ensemble Methods: Foundations and Algorithms*, 1st edn. (Chapman &Hall/CRC, Boca Raton, 2012)

Index

F. Torres-Guerrero et al. (eds.), *2nd EAI International Conference on Smart Technology*, EAI/Springer Innovations in Communication and Computing,
https://doi.org/10.1007/978-3-031-07670-1